Agent-Based Social Systems

Volume 7

Editor in Chief:

Hiroshi Deguchi, Yokohama, Japan

Series Editors:

Shu-Heng Chen, Taiwan, ROC
Claudio Cioffi-Revilla, USA
Nigel Gilbert, UK
Hajime Kita, Japan
Takao Terano, Japan

For other titles published in this series, go to
www.springer.com/series/7188

ABSS–Agent-Based Social Systems

This series is intended to further the creation of the science of agent-based social systems, a field that is establishing itself as a transdisciplinary and cross-cultural science. The series will cover a broad spectrum of sciences, such as social systems theory, sociology, business administration, management information science, organization science, computational mathematical organization theory, economics, evolutionary economics, international political science, jurisprudence, policy science, socioinformation studies, cognitive science, artificial intelligence, complex adaptive systems theory, philosophy of science, and other related disciplines.

The series will provide a systematic study of the various new cross-cultural arenas of the human sciences. Such an approach has been successfully tried several times in the history of the modern science of humanities and systems and has helped to create such important conceptual frameworks and theories as cybernetics, synergetics, general systems theory, cognitive science, and complex adaptive systems.

We want to create a conceptual framework and design theory for socioeconomic systems of the twenty-first century in a cross-cultural and transdisciplinary context. For this purpose we plan to take an agent-based approach. Developed over the last decade, agent-based modeling is a new trend within the social sciences and is a child of the modern sciences of humanities and systems. In this eries the term "agent-based" is used across a broad spectrum that includes not only the classical usage of the normative and rational agent but also an interpretive and subjective agent. We seek the antinomy of the macro and micro, subjective and rational, functional and structural, bottom-up and top-down, global and local, and structure and agency within the social sciences. Agent-based modeling includes both sides of these opposites. "Agent" is our grounding for modeling; simulation, theory, and real-world grounding are also required.

As an approach, agent-based simulation is an important tool for the new experimental fields of the social sciences; it can be used to provide explanations and decision support for real-world problems, and its theories include both conceptual and mathematical ones. A conceptual approach is vital for creating new frameworks of the worldview, and the mathematical approach is essential to clarify the logical structure of any new framework or model. Exploration of several different ways of real-world grounding is required for this approach. Other issues to be considered in the series include the systems design of this century's global and local socioeconomic systems.

Keiki Takadama · Claudio Cioffi-Revilla
Guillaume Deffuant
Editors

Simulating Interacting Agents and Social Phenomena

The Second World Congress

Springer

Editors
Keiki Takadama
Associate Professor
Department of Informatics
and Engineering
The University of Electro-Communications
1-5-1 Chofugaoka, Chofu
Tokyo 182-8585
Japan
keiki@hc.uec.ac.jp

Claudio Cioffi-Revilla
Professor
Center for Social Complexity
George Mason University
4400 University Drive
MSN 6B2, Fairfax
Virginia 22030
U.S.A.
ccioffi@gmu.edu

Guillaume Deffuant
Head of LISC (Laboratoire d'Ingénierie
des Systèmes Complexes), Cemagref
24 avenue des Landais
B.P. 50085, 63172 Aubière
Cedex 1, France
guillaume.deffuant@cemagref.fr

ISSN 1861-0803
ISBN 978-4-431-54081-6 ISBN 978-4-431-99781-8 (eBook)
DOI 10.1007/978-4-431-99781-8
Springer Tokyo Dordrecht Heidelberg London New York

Printed on acid-free paper

Springer is part of Springer Science+Business Media (www.springer.com)

Preface

We are pleased to publish *Simulating Interacting Agents and Social Phenomena: The Second World Congress* as the post-proceedings of the Second World Congress on Social Simulation (WCSS '08) held at George Mason University in the United States, 14–17 July 2008.

As the papers selected for this volume testify, computational social science continues to grow as a dynamic scientific interdisciplinary field driven by the medium of computing. The innovations in the field include new concepts, theories, models, and methodologies across a broad variety of approaches, including social simulation, complexity-theoretic research, social network analysis, automated information extraction algorithms, and related information-processing approaches to understanding and explaining human and social dynamics. Moreover, the contributions span the five classical disciplines of social science—anthropology, economics, sociology, political science, and social psychology—because no social science today lacks significant advances brought about by the computational approach.

In addition, the presence of computational social science and social simulation is also contributing new knowledge in interdisciplinary areas, such as management science, organization theory, social geography, communication, archaeology, and the policy sciences. Quite often, what unites these diverse investigations is precisely the computational approach, typified but not exclusively represented by social simulation and agent-based modeling.

From the perspective of just how deep computational approaches have penetrated the investigation of social phenomena at all levels, it is significant to note that practically all areas or subfields of specialization in social science by now have witnessed some computational contributions. For example, as demonstrated by chapters in this volume, economics is witnessing computational social simulation contributions in specialized areas as diverse as finance, markets, and microeconomics, as well as international trade and development. Similarly, in political science computational contributions now cover comparative politics as well as international relations, among other subfields. In anthropology, both cultural anthropology and anthropological archaeology are also witnessing significant computational advances. In other words, the penetration of computational approaches is deep and lasting, to the benefit of scientific progress.

The papers included in this volume were selected by a rigorous process of double peer review. First, all papers accepted for presentation at the Second World

Congress on Social Simulation were selected by a peer-review system whereby the selected papers were revised based on reviewers' feedback. Second, following the congress, an additional review of selected papers was conducted, resulting in further revisions that finally led to the chapters included in this volume. We are very grateful to the members of the Program Committee who dedicated time and talent to this rigorous process.

The Second World Congress on Social Simulation held at George Mason University, near Washington DC, followed the excellent international tradition established by the First World Congress held at Kyoto University two years earlier, by hosting scholars from many countries. These included Australia, Brazil, Ecuador, France, Germany, Italy, Japan, Poland, Rumania, the Netherlands, the Russian Federation, Spain, Sweden, Switzerland, Taiwan, the United Kingdom, and the United States, among many others. This vast international participation was due to the hard work of the regional associations that have jointly sponsored the World Congress series, namely, the Pacific Area Association for Agent-based Social Systems Science (PAAA), the European Social Simulation Association (ESSA), and the North American Association for Computational Social and Organizational Sciences (NAACSOS), which has now been reorganized as the Computational Social Science Society (CSSS, or C-Triple-S). Already the next international conference, the Third World Congress on Social Simulation (WCSS '10) at Kaasel University, Germany, 6–9 September 2010, is being organized as this volume is being published.

As editors, we wish to express deep appreciation to Professor Hiroshi Deguchi, Tokyo Institute of Technology, and the editorial staff at Springer Japan, who facilitated the publication of this volume in the ABSS Series. Besides the many academic faculty and professionals who contributed to the success of the Second World Congress, we are grateful to the many students from around the world who participated with papers and posters. Students added a vibrant and meritorious representation of the next generation of future computational social scientists and social simulation experts. Finally, a word of sincere gratitude to the administrative staff that supported the local committee at the Mason Center for Social Complexity, especially Christina Bishop and Beth Gronke, as well as the conference web site developer, Nicolas Dumoulin, from Laboratoire d'Ingénierie des Systèmes Complexes (Cemagref, France), who worked many hours beyond the call of duty to ensure an excellent international congress.

Editors

Keiki Takadama (PAAA)
Claudio Cioffi-Revilla (CSSS)
Guillaume Deffaunt (ESSA)

Contents

WCSS'08 Organization

Chairs

Claudio Cioffi-Revilla, George Mason University, USA
Guillaume Deffuant, Cemagref, France

Program Committee Co-Chairs

David Hales, University of Bologna, Italy
David Sallach, Argonne National Laboratory, USA
Keiki Takadama, The University of Electro-Communications, Japan

Program Committee

North America (NAACSOS):

Steven Bankes, Evolving Logic
Kathleen Carley, Carnegie Mellon University
Daniel Diermeier, Northwestern University
Deborah Vakas Duong, US-OSD Simulation Analysis Center
William Griffin, Arizona State University
Stephen Guerin, Redfish Group
Marco Janssen, Arizona State University
Charles Macal, Argonne National Laboratory
Michael North, Argonne National Laboratory
Jonathan Ozik, Argonne National Laboratory
Dawn C. Parker, George Mason University
Michael Prietula, Emory University
William Rand, Northwestern University
Dwight Read, University of California at Los Angeles
Robert Reynolds, Wayne State University
Fabio Rojas, Indiana University
Kevin Ruby, University of Chicago
Keith Sawyer, Washington University

Asia and Pacific (PAAA):

Shu-Heng Chen, National Chengchi University, Taiwan
Sung-Bae Cho, Yonsei University, Korea

Norman Foo, University of New South Wales, Australia
Scott Heckbert, Commonwealth Scientific and Industrial Research Organisation, Australia
Toshiyuki Kaneda, Nagoya Institute of Technology, Japan
Toshiji Kawagoe, Future University-Hakodate, Japan
Hajime Kita, Kyoto University, Japan
Yusuke Koyama, Tokyo Institute of Technology, Japan
Deddy Priatmodjo Kusrindartoto, Institut Teknologi Bandung, Indonesia
Hiroyuki Matsui, Kyoto University, Japan
Yosihiro Nakajima, Osaka City University, Japan
Philippa Pattison, The University of Melbourne, Australia
Utomo Sarjono Putro, Institut Teknologi Bandung, Indonesia
Hiroshi Sato, National Defense Academy, Japan
Shingo Takahashi, Waseda University, Japan
Takao Terano, Tokyo Institute of Technology, Japan
Sun-Chung Wang, Academia Sinica, Taiwan
David W. K. Yeung, Hong Kong Baptist University and St. Petersburg State University, China
Suiping Zhou, Nanyang Technological University, Singapore

Europe (ESSA):
Frederic Amblard, University of Toulouse, France
Luis Antunes, University of Lisbon, Portugal
Olivier Barreteau, Cemagref, France
Francois Bousquet, CIRAD, France
David Chavalarias, Ecole Polytechnique, France
Helder Coelho, University of Lisbon, Portugal
Rosaria Conte, CNR Rome, Italy
Nuno David, ISCTE, Lisbon, Portugal
Alexis Drogoul, IRD, France
Bruce Edmonds, CPM Manchester, UK
Nigel Gilbert, University of Surrey, UK
Nick Gotts, The Macaulay Institute, Aberdeen, UK
Dirk Helbing, ETH Zurich, Switzerland
Luis R. Izquierdo, University of Burgos, Spain
Wander Jager, University of Groningen, Netherlands
Juergen Kluever, University of Essen, Germany
Scott Moss, CPM Manchester, UK
Mario Paolucci, CNR Rome, Italy
Juliette Rouchier, GREQAM, CNRS, France
Frank Schweitzer, ETH Zurich, Switzerland
Jaime Sichman, University of Sao Paulo, Brazil
Flaminio Squazzoni, University of Brescia, Italy
Klaus G. Troitzsch, University of Koblenz, Germany

Local Organisation Committee

Robert Axtell
Dawn C. Parker
Maksim Tsvetovat
Christina Bishop
Beth Groenke

Part I
Norms, Diffusion and Social Networks

A Classification of Normative Architectures

Martin Neumann

Abstract In this paper, a comparative analysis of selected cases of normative agent architectures is undertaken. A classification scheme is introduced that represents crucial decisions of which designers of normative agents have to be aware. Preconditions are the theoretical background on which the author relies and whether single or social embedded agents are considered. Designers then have to specify the conceptual realization of norms, to decide whether norm dynamics are to be considered and how to deal with possible conflicts.

Keywords Agent architectures • Classification • Norms

1 Introduction

In the past decade, the study of norms has been one of the most active research fields within multi-agent based simulation. Norms are a useful tool and a crucial theoretical concept in agent-based modelling: the motivations to include the notion of social norms in the design of multi-agent systems range from technical problems in the application of multi-agent systems [16, 30] to theoretical interest in the foundations of Distributed Artificial Intelligence [9, 15] and the 'wheels of social order' [13].

Conceptually, there exist two different approaches to investigate norms by the means of multi-agent systems [22]: On the one hand, norms are realised as an emergent property in evolutionary game theoretical models. Examples of such an approach are the models of Axelrod [2], Skyrms [28, 29] or Young [33, 34]. These models focus on norm *dynamics*. An overview of this research field can be found in [18]. On the other hand, norms are implemented in cognitive agents of the AI tradition. The motivation to explicitly include the cognitive dimension of norms

M. Neumann (✉)
RWTH Aachen University, Institute for Sociology, Eilfschornsteinstr. 7, 52062 Aachen, Germany
e-mail: mneumann@soziologie.rwth-aachen.de

K. Takadama et al. (eds.), *Simulating Interacting Agents and Social Phenomena: The Second World Congress*, Agent-Based Social Systems 7,
DOI 10.1007/978-4-431-99781-8_1, © Springer 2010

in the design of AI agents is to differentiate norms from mere coercion [1] as well as simple conventions [12]. The conceptual difference between the two approaches can be illustrated by the difference between understanding and prediction, highlighted in the philosophy of the humanities: while models of evolutionary game theory seek to explain the emergence of norms by some very general assumptions, systems of cognitive agents can be regarded as a hermeneutics of norms: they are a detailed description of the intuitions about norms and aim to provide a platform to investigate the implications of these intuitions in simulation experiments. The objective of this article is to focus on the challenges imposed by the inclusion of norms in the agent's design. Therefore a review of agent architectures with an explicit normative layer in the agent's design is undertaken. Hence, since in game theoretical models norms are not a built-in property of the agent's architecture, it concentrates on *AI architectures*.

Technically, norms are implemented in multi-agent systems to enhance the stability of the system. In particular, norms can be used to coordinate decentralised open systems with several designers [16]. This becomes relevant in applications such as e-commerce and electronic institutions [20, 30]. In such cases, agents may work on behalf of different users with potentially conflicting interests. While the technical challenge is to combine agent autonomy with social predictability [7, 31], the theoretical dimension behind these applications is the question of trust among strangers.

However, in the design of normative agents, coordination problems and potential conflicts among agents' different internal informational and motivational components, such as goals, desires, intentions and obligations, have to be resolved [7, 8]. Traditional, formal treatments of this problem follow the BDI model, as proposed by Rao and Georgeff [24]. However, specifying the relation between components remains an open research field [15].

Until now, the inclusion of norms remains a major challenge in the design of multi-agent systems. This is indicated by numerous conceptual considerations about the architecture of normative agents. In fact, the number of conceptually oriented articles on the architecture of normative agents exceeds the number of existing models. Typically, norms in concrete models are less sophisticated than concepts proposed in formal architectures [14]. However, focusing on implementation before having achieved a proper understanding of formal architectures risks loosing key norm-related intuitions [15]. The development of architectures is a kind of *requirement analysis*: it specifies the essential components of normative agents.

The number of already existing architectures suggests that we undertake a meta-analysis of the proposed requirements: summarising the different approaches to normative agent architectures in order to achieve a classification of design decisions of which modellers must be aware. This should give designers an overview of options to match their own purposes with the current state of the art. Moreover, this classification provides a list of standards agreed upon in modelling normative agents. This matches current attempts to develop a standard protocol [17, 25] to enhance communication and the intersubjective validity of agent-based modelling. The architectures are evaluated against five design decisions that are undertaken implicitly by the authors. This includes the choice of the formal concept in which norms are formulated, here represented as a scale from logic to decision theory. This is the very first decision that has to be undertaken, simply to get started. The next design decision is whether one single agent is investigated or a social level, i.e. a population of (interacting) agents, is considered. Moreover, implicitly the architectures make statements about their philosophical orientation: namely, while some are formulated in terms of utilities, others rely on deontic concepts of norms. This reflects the difference between consequentialistic and deontic foundations of moral philosophy. Next, it is investigated whether and what kind of dynamics is included in the architectures and finally if and what kind of conflicts are captured. It is not possible to prove that this is an exhaustive list of all possible design decisions. However, it can be revealed within the different architectures and allows to discriminate between them. Moreover, it has been evaluated by an expert discussion at the second world congress of social simulation. Thus there is evidence that it is empirically reliable.

However, it is not possible to include all existing architectures within the scope of this paper. This review will concentrate on a representative sample of case studies, covering the range of possible options, for closer inspection. This allows us to detect the crucial design decisions of which modellers of normative agents have to be aware. Section 2 gives an overview of 14 selected cases. In Sect. 3, the classification of design decisions will be elaborated, including details on how the cases fit into this scheme. Section 4 closes the paper with concluding remarks.

2 The Selected Cases

- On the Immergence of Norms [1]. Hereinafter referred to as *Immergence*. This paper investigates the process of norm innovation. The behaviour of an agent may be interpreted by an observing agent as normative if it is salient in the observer's normative board. Thus, norm instantiation is regarded as an inter-agent process.

  Implementation/prototype: yes (described in [21])

- Norm Governed Multiagent Systems [4] *(Norm Governed)*. This paper differentiates between three types of agents: agents who are the subject of norms, so-called defender agents, who are responsible for norm control, and a normative authority that has legislative power and that monitors defender agents.

  Implementation/prototype: no

- An architecture of a normative System [5] *(Architecture)*. The authors rely on John Searle's notion of institutional facts (so-called 'counts-as' conditionals) to represent social reality in the agent architecture. A norms base and a 'counts-as' component transforms brute facts into obligations and permissions.

  Implementation/prototype: no

- The BOID Architecture [7] *(BOID)*. The Belief-Obligation-Intentions-Desire (BOID) Architecture is the classical approach to represent norms in agent architectures. Obligations are added to the BDI Architecture to represent social norms while preserving the agent's autonomy. Principles of the resolution of conflicts between the different components are investigated in the paper.

  Implementation/prototype: yes

- Norms in Artificial Decision Making [6] *(Artificial Decision)*. This paper proposes the use of super-soft decision theory to characterise real-time decision-making in the presence of risk and uncertainty. Moreover, agents can communicate with a normative decision module to act in accordance with social demands. Norms act as global constraints on individual behaviour.

  Implementation/prototype: no

- Deliberative normative agents [10] *(Deliberative)*. This paper explores the principles of deliberative normative reasoning. Agents are able to receive information about norms and society. The data is processed in a multi-level cognitive architecture. On this basis, norms can be adopted and used as meta-goals in the agent decision process.

  Implementation/prototype: no

- From conventions to prescriptions [12] *(Prescriptions)*. A conventionalist (in rational philosophy) and a prescriptive (in philosophy of law) perspective on norms is distinguished in this paper. A logical framework is introduced to preserve a weak intuition of the prescriptive perspective which is capable of integrating the conventionalist intuition. The notion of a normative authority is therefore abandoned.

  Implementation/prototype: no

- From Social Monitoring to Normative Influence [14] *(Influence)*. This paper argues that imitation is not sufficient to establish a cognitive representation of norms in an agent. Agents infer abstract standards from observed behaviour. This allows for normative reasoning and normative influence in accepting (or defeating) and defending norms.

  Implementation/prototype: no

- From desires, obligations and norms to goals [15] *(Desires)*. This paper investigates the relations and possible conflicts between different components in an agent's decision process. The decision-making process of so-called 'B-doing agents' is designed as a two-stage process, including norms as desires of society. The authors differentiate between abstract norms and concrete obligations.

  Implementation/prototype: no

- Norm-oriented programming of electronic institutions [16] *(Institutions)*. In this paper, norms are introduced as constraints to regulate the rules of interaction between agents in situations such as a Dutch auction protocol. These are regulated by an electronic institution (virtual auctioneer) with an explicit normative layer.

Implementation/prototype: yes

- An architecture for autonomous normative agents [20] *(Auto Norms)*. This paper explores the process of adopting or rejecting a normative goal in the BDI framework. Agents must recognise themselves as addressees of norms and must evaluate whether a normative goal has a higher or lower priority than those hindered by punishment for violating the norm.

  Implementation/prototype: no

- Normative KGP Agents [26] *(KGP)*. The authors of this paper extend their concept of knowledge, goals and plan (KGP) agents by including norms based on the roles played by the agents. For this reason, the knowledge base KB of agents is upgraded by KB_{soc}, which caters for normative reasoning, and KB_{rev}, which resolves conflicts between personal and social goals.

  Implementation/prototype: yes

- On the synthesis of useful social laws for artificial agent societies [27] *(Synthesis)*. The authors propose building social laws into the action representation to guarantee the successful coexistence of multiple programs (i.e. agents) and programmers. Norms are constraints on individual freedom. The authors investigate the problem of automatically deriving social laws that enable the execution of each agent's action plans in the agent system.

  Implementation/prototype: no

- Norms in Multi-Agent Systems: From theory to practice [30] *(Theory & Practice)*. This paper provides a framework for the normative regulation of electronic institutions. Norms are instantiated and controlled by a central institution, which must consist of a means to detect norm violation and a means to sanction norm violators and repair the system.

  Implementation/prototype: no

3 A Classification of Design Decisions

3.1 Design Decision 1: The Scale from Logic to Decision Theory

First and foremost, designers of normative agent architectures have to communicate their intuitions about norms in some way. Hence they have to decide how to represent the structure of the architecture. Three theoretical backgrounds can be found in the literature: reliance on deontic logic, reliance on – some form of – decision theory, or extension of the BDI approach. Deontic logic has been proposed as a formalism for reasoning about obligation in the philosophical literature [32]. This relies on the assumption that verbs such as ‘ought’ or ‘should’ can be reconstructed as modalities in the possible world framework. Since the construction of agents is in need of formal ‘if then’ relations, it is plausible to apply deontic logic to the formalisation of normative agents. Similar to the logical approach, agent representations in the Belief Desire (BD) framework provide a formal treatment of defining agents. However, while deontic logic investigates implications between obligations, approaches following the BD approach concentrate on the relations between *different* components, such as beliefs, desires and – in the case of normative agents – obligations [3]. The logical relations between these components are not yet fully understood [15]. For instance, it is a common practice to explicitly differentiate obligations and permissions, even though logically both modal operators could be transformed into one another. Moreover, the role of sanctions is typically not considered in logical approaches [3]. However, it is crucial to evaluate the consequences of sanctions to obtain rational decisions. Thus, even though the BD approach investigates logical relations, it also shares distinct features with decision theory. The central concept of decision theory is the notion of utility, which goes beyond the scope of logic. Utility is central to the agents’ decision-making process. It is thus another option to rely on decision theory for the agent design. However, only one paper relies explicitly on ‘supersoft’ decision theory [6]. This theoretical framework is usually only implicitly applied. Here, reference will be made to the notion of decision theory if agents make decisions that are not based on logical implications or an explicit BD structure (Table 1).

In the following, this classification will be illustrated by one example from each case: prescriptions provides a striking example of a logical approach. After an exposition of the problem, it concentrates on logical definitions of a normative formalism that allows to overcome this problem. Individual agents are only considered as variables. The BOID architecture provides a straightforward extension of the BDI approach by including a further component in the cognitive agent architecture. However, the decision whether to subsume an architecture under the label decision theory is more subtle. Yet influence, for instance, is labelled as decision theoretic since the main focus of the paper is to analyse the decision process when to comply with norms in a social environment rather than to provide a detailed description of the cognitive architecture of a single agent.

Table 1 Theoretical background

Logic	BD formalisation	Decision theory
Prescriptions	KGP	Immergence
Synthesis	BOID	Influence
Institutions	Desires	Artificial Decision
Theory & Practice	Auto Norms	
	Norm Governed	
	Architecture	
	Deliberative	

3.2 *Design Decision 2: Single or Social Agents?*

Norms are described as the 'social burden' [16] or 'desires of the society' [15]. They become relevant when agents interact with other agents. However, norms are also a feature of individual agents, regulating individual behaviour. Norms are a link between the individual and society. Designers of normative agents therefore have to make a crucial decision whether to concentrate on the cognitive architecture of a single agent or whether to take the agents' interaction structure into account. While it allows for a more differentiated architecture of the intra-agent processes to concentrate on a single agent, inclusion of the interaction processes between agents allows for the investigation of the dynamic character of norms. A compromise between these two alternatives is to represent society in an indirect manner: by modelling society only implicitly as a mental representation of a single agent (Table 2).

In the following, this classification will be illustrated. The cases of social agents, i.e. populations of interacting agents, and single agents are rather self-evident. Mental representation can be explained by the example of 'Deliberative': The paper consists of a detailed description of a highly complex cognitive architecture of an individual agent. This architecture includes (next to norm components) also a component to store information about the society. The information has to be implemented externally because no other agents are actually present. No interaction is described. Hence the society is only implicitly present in the agent's mind.

Moreover, the representation of social complexity is highly different in different architectures. This refers to the notion of social roles. If and how social roles are specified is of crucial relevance for the concept of norms. One central distinction is between two central roles related to norms: addressees of norms who are obliged to fulfil the norms and beneficiaries of norms. However, roles can be further specified: for instance, in the KGP architecture roles are assigned to agents, which are initiated and terminated by an event calculus; e.g. assign(a, t_w(chelsea, 9, 7)) means that agent a plays the role of a traffic warden in Chelsea between the times 9 a.m. and 5 p.m. This implies that agent a is addressee of different obligations between 9 a.m. and 5 p.m. (i.e. its working duties) than at other times.

Interestingly, the introduction of social roles is not restricted to architectures of systems of agents. While the KGP architecture investigates only the architecture of a single agent, it has in fact the most differentiated concept of roles. However, this

Table 2 Social embedding

Social agents	Mental representation	Single agent
Norm Governed	Deliberative	BOID
Influence	Prescriptions	KGP
Immergence	Theory & Practice	Architecture
	Artificial Decisions	Desires
	Institutions	Synthesis
		Auto Norms

is an exception. Roles that exceed the notion of addressees and beneficiaries are explicitly introduced in Norm Governed, Theory & Practice and KGP. However, only KGP is restricted to a single agent.

3.3 Design Decision 3: Concepts of Norms

The above-mentioned design decisions refer to *preconditions* for modelling norms. The following distinctions specify the concept of norms used in the different architectures.

In the philosophical literature [19] it is traditionally distinguished between deontic and consequentialist concepts of norms. Deontic concepts of norms emphasise that norms are in itself a reason for action. A moral judgement of an action is guided by the reasons for the actions. People are prescribed to follow norms regardless of the consequences. Presumably, the most prominent example of such a deontic concept is Kant's categorical imperative. On the other hand, consequentialist norm concepts judge actions not by its reasons but by its consequences. Utilitarism is the most prominent example of this a philosophical position. Hence, the design of normative agents cannot avoid to express a philosophical stance (Table 3).

In the following, this classification will be illustrated by one example from each case: no philosophical theories are mentioned in the papers. However, the implementation of a separate obligations component in BOID provides a good example of a deontic point of view. While the agent is able to undertake means-ends calculations to determine the actions in order to realise a certain intention, the norms do not derive from these calculations. They are given by an external instance (in this case: the programmer). They provide in itself a reason for action. This is the point of view of deontic moral philosophy. Indeed, normative AI architectures are biased towards this position because they do not simply regard norms as a behaviour regularity as, for instance, game theoretic models. Rather they treat norms as a cognitive object. Nevertheless, there are great differences with respect to the 'selfishness' of the agents. In Norm Governed, for instance, the agents obey norms only if they believe to be observed by a punishing agent. Here, norms are *not in itself* a reason for action. This comes close to game

Table 3 Philosophical background

Deontic	Consequentialistic
BOID	Norm Governed
Deliberative	Influence
Synthesis	Immergence
Prescriptions	Institutions
Artificial Decisions	Auto Norms
Theory & Practice	Desires
Architecture	
KGP	

theoretic explanations of norm compliance (e.g. [2]). Agents follow norms when they are faced with the *consequences* of norm violation.

However, beside the general philosophical approach, the character of norms has to be in some way specified computationally. The existing approaches can be regarded as a hierarchy of increasingly sophisticated accounts. The simplest and most straightforward way is to regard norms as mere constraints on the behaviour of individual agents. For example, the norm to drive on the right-hand side of the road restricts individual freedom. In this case, norms need not necessarily be recognised as such. They can be implemented off-line or can emerge in interaction processes. It follows that it is not possible to distinguish the norm from the normal. More sophisticated accounts treat norms as mental objects [10, 11]. This allows for deliberation about norms and, in particular, for the conscious violation of norms. Norms intervene in the process of goal generation, which might – or might not – lead to the revision of existing personal goals and the formation of normative goals. However, two approaches also have to be distinguished in this case: a number of accounts (such as the BOID architecture) rely on the notion of obligations. Obligations are explicit prescriptions that are always conditional to specific circumstances. One example of an obligation is not being permitted to smoke in restaurants. Agents may face several obligations that may contradict one another. For this reason, some authors differentiate between norms and obligations. Norms are regarded as more stable and abstract concepts than mere obligations [14, 15]. One example of such an abstract norm is 'being altruistic': further inference processes are needed for the formation of concrete goals from this abstract norm. Such abstract norms are timelessly given and not context specific (Table 4).

In the following, this classification will be illustrated by one example from each case: in Synthesis it is simply an implemented rule that robots drive on the right side. Normative KGP agents, in contrast, have several components for several tasks (e.g. for action planning or goal decision). These components include a notion of the social role of the agent and the obligations, permissions and prohibitions that are associated with this role. Moreover, the agents possess a component KB_{rev} that represents the agents preference policy towards revising incompatible personal, reactive and social goals. This allows for more freedom of action than to implement norms as a mere constraint. The most characteristic example of an abstract conception of norms is Influence. For the purpose to

Table 4 Norm concepts

Constraints	Obligations	Abstract concepts
Institutions	KGP	Desires
Artificial Decision	BOID	Immergence
Synthesis	Architecture	Influence
	Auto Norms	
	Norm Governed	
	Prescriptions	
	Theory & Practice	
	Deliberative	

Table 5 Norm dynamics

Static	Norm spreading	Norm innovation
Artificial Decision	Influence	Immergence
Synthesis	Prescriptions	
KGP		
BOID		
Architecture		
Auto Norms		
Theory & Practice		
Institutions		
Deliberative		
Norm Governed		
Desires		

develop a notion of normative influence rather than to equate social learning with imitation, it introduces abstract (normative or moral) standards as a criterion to evaluate others.

3.4 Design Decision 4: Static or Dynamic Norms?

Obviously, norms may change in the course of time. One further important difference is thus whether norms are a static or dynamic concept. In the latter case, two different cases can be distinguished: the innovation of new norms and a changing scope of validity of the norm (Table 5).

These categories are rather evident. In the BOID case, for instance, obligations are programmed-in. They cannot change in the course of a simulation. Influence, on the other hand, is concerned with the process of convincing other agents to comply with a norm. This is an instance of norm spreading. However, new norms emerge only in the case of Immergence. Here agents observe the behaviour of other agents and decide whether they regard this behaviour as normative or not. If an accidental behaviour is regarded as normative, this can give raise to the emergence of a new norm.

3.5 Design Decision 5: Norm Conflicts

Technically, norms could be simply implemented into the agent's desires component. In fact, this has been the suggestion in the early paper 'on the synthesis of useful social laws for artificial agent societies' by Shoham and Tennenholtz [27]. However, the intuition to introduce a separate norms component into the agents' design is to ensure an agent's autonomy: by explicitly separating individual and social desires, it is possible that the agent can deliberate over which component has priority.

The intuition behind this design decision is to distinguish the agents' goals and desires from the demands of the society. This implies that conflicts might arise between the components. Hence, conflicts are central for this intuition of norms. Yet the concept of conflicts is not as unequivocal as it might appear at first sight. The question appears where to identify the source of conflicts. Conflicts may arise between agents, between the agents' goals and social norms, and there might even be contradicting social norms. While norm invocation can be regarded as a conflicts between agents, conflicts between different components arise when social norms contradict individual desires, e.g. the desire to smoke in a non-smoking area. Norms can contradict each other especially if they are prescribed by different institutions or normative authorities. In fact, all these intuitions can be identified in the various architectures (Table 6).

In the following, this classification will be illustrated by one example from each case: the first case is simple. E.g. in Synthesis the implementation of the rule 'drive on the right hand side' is one action rule next to others. Contradictions would cause an error. In Institutions conflicts between agents arise because it deals with the regulation of electronic institutions. This is discussed by the example of a Dutch auction. The (electronic) auctioneer is necessary to balance the conflicting interests of different agents. BOID is a straightforward example of a case that allows for conflicts between goals and norms because they are stored in different components. Theory & practice investigates different ways to handle conflicting legal norms, i.e. conflicts between different norms. It suggests e.g. context specification, or an ordering from abstract to more concrete versions of the norms for the purpose of conflict resolution.

Indeed, the consideration of conflicts calls for some kind of conflict resolution, even though in some cases this is not considered explicitly. Other architectures, however, develop highly sophisticated techniques. A straightforward idea is the maximisation of expected utility [20]. This implies a consequentialist conception of norms.

The conflict resolution of KGP, Desires and the BOID architecture can be subsumed under the general idea of priority ordering between the different components, i.e. beliefs, obligations, intentions or desires. The concrete realisation, however, is quite different from case to case. In particular in the BOID architecture a demanding logical formalism is developed, including a feedback loop and a consideration of all consequences of an action to detect possible

Table 6 Conflicts

Not considered	Between agents	Between goals and norms	Between norms
Prescriptions	Institutions	Artificial Decision	Desires
Synthesis	Influence	KGP	Immergence
Architecture		BOID	Theory & practice
		Auto Norms	
		Deliberative	
		Norm Governed	

contradictions. A priority ordering between components allows to differentiate between kinds of agents such as selfish or socially responsible agents. For instance, in the case of selfish agents, elements of the desires components overrule elements of the obligations component if they contradict each other. Thus this kind of conflict resolution is restricted to contradictions between goals and norms.

To handle with conflicts between different social obligations other concepts are necessary. The architecture developed in Immergence [1] introduces norm innovation to integrate conflicting norms. This implies that agents have to be able to autonomously develop new semantic concepts. At the example of the code of law, different ways to formalise exception rules are discussed in Theory & Practice [30]. As it is known from default reasoning, the lower level of abstraction specifies exceptions to default.

4 Conclusion

In order to obtain a final insight into the main principles of the design of normative agent architectures, an overview of the cases is given (Table 7):

What can be learned from the comparative look on different architectures of normative agents?

- First and foremost, the variability of concepts is striking: Typically, norms are introduced in the AI literature without much discussion. It is presumed that the intuition about norms is clear and straightforward. However, the review shows that this is not the case. Once the designer develops an architecture, she is forced

Table 7 Overview of the architectures

	Backgr.	Soc. emb.	Phil. back.	Concepts	Dynamics	Conflicts
Prescriptions	Logic	Repres.	Deontic	Obligation	Spreading	Not cons.
Institutions	Logic	Repres.	Consequ.	Constraint	Static	Bet. agents
Synthesis	Logic	Single	Deontic	Constraint	Static	Not cons.
Theory & Prac.	Logic	Repres.	Deontic	Obligation	Static	Bet. norm
Norm Governed	BD	Social	Consequ.	Obligation	Static	Goal/norm
Architecture	BD	Single	Deontic	Obligation	Static	Not cons.
BOID	BD	Single	Deontic	Obligation	Static	Goal/norm
Deliberative	BD	Repres.	Deontic	Obligation	Static	Goal/norm
Desires	BD	Single	Consequ.	Concept	Static	Bet. norm
Auto Norms	BD	Single	Consequ.	Obligation	Static	Goal/norm
KGP	BD	Single	Deontic	Obligation	Static	Goal/norm
Immergence	DT	Social	Consequ.	Concept	Innovation	Bet. norm
Art. Decision	DT	Repres.	Deontic	Constraint	Static	Goal/norm
Influence	DT	Social	Consequ.	Concept	Spreading	Bet. agents

DT Decision Theory, *BD* Belief-Desire

to provide a detailed description of the concept of norms. Then, however, it becomes obvious that very different intuitions exist.
- A short look at the above table shows that the design decisions have no taxonomic structure, as in the biological taxonomy of species. To a large degree, design decisions are *independent* of the decisions taken at other stages. For instance, a Decision Theoretic approach is compatible with socially embedded agents as well as with a mental representation of society. The latter, however, can also be achieved using the Belief-Desire or a logical approach. Also the sophisticated view on norms as an abstract concept has been realised with both Decision Theory and BD formalisations. Even a reliance on the theoretical background of deontic logic is compatible with a consequentialist concept of norms.
- However, the table shows that most architectures apply a static approach to norms. A dynamic view has not been fully explored. However, this is the particular strength of evolutionary game theory. Models applying a very simple concept of agents are able to provide insights into the dynamics of norms and how they are established. To catch up with the state of the art in evolutionary game theory with cognitively rich agents, the research agenda should focus on norm innovation. Only the paper on 'Immergence' points in this direction.

Architectures of cognitive agents aim at a hermeneutics of norms by providing a detailed description of the intuition of norms. What would be the most detailed description of the (presumably) most common intuition about norms? To stimulate further discussion, in the following some suggestions will be made along the lines of the classification of design decisions:

- With regard to the *theoretical background* it can be argued that architectures following the BDI approach (such as BOID architectures) provide the most detailed description of cognitive processes.
- Obviously, a realistic representation of human agency should entail a *social embedding*. A detailed description should provide both a mental representation of society as well as other agents. The society and its mental representation need not be identical. In particular, agents should be embedded in a structured social environment. Human society entails social roles and social stratification.
- It is impossible to provide an unequivocal advise with respect to the *philosophical background*. The dispute between deontic and consequentialist foundations of ethics is as old as philosophy. However, everyday normative reasoning entails reference to some mental concepts of norms.
- The most ambitious computational *concept* of norms is to formulate norms as an abstract concept that is more stable than concrete obligations and permissions in specific circumstances.
- Obviously, social norms and values are subject of change. Norm *dynamics* is a feature of real societies.
- All intuitions formulated by the different types of *conflicts* appear to be very plausible. Sanctioning, or norm invocation is a conflict between agents. It is also

a common experience that individual goals can contradict social prescriptions as well as dilemma situations in which the individual agent has to decide between contradicting norms. Presumably, this is the most demanding instance of normative reasoning. Thus it appears to be desirable that all kinds of conflicts could be handled by a comprehensive architecture.

No single architecture can be rendered as formulating the one and only most sophisticated concept of norms: 'Immergence' includes social embedding, an abstract concept and innovation of norms. 'Influence' describes a comparable approach. 'KGP', 'Theory & Practice' and 'Norm Governed' introduce the notion of social roles. A hypothetical, theoretical most ambitious architecture would be a combination of these. It would be build upon the BDI approach and would entail social embedding, both on a physical level and as a mental representation. The social level should include a dynamics of norms which should be recognized by the agents at the level of their mental representation of the society. It would include a concept of abstract norms and a reasoning process that would enable to evaluate between competing norms. This would include that the agents possesses a concept of deontics as well as some kind of utility calculations. This could be a filtering mechanism: first, the agent might filter out those norms which are in contrast to abstract 'moral' values and then the agents might select from the remaining norms those which are most consistent with their own desires (Table 8).

However, for practical purposes careful considerations of what is needed for concrete applications in the respective target systems are indispensable. For instance, to regulate electronic institutions, representations of cognitive processes are not needed, while social embedding is essential. In particular, the features of the target system have to be closely examined. They might exhibit very different characteristics than the generating individual agents: economic systems, for instance, emerge without a doubt from complex, deliberate individual agents. However, the emergent properties of the economic systems might not be in need of such agents with highly sophisticated cognitive capacities. On the contrary, they might at best be explained by simple particle dynamics [23]. 'Synthesis' provides an example of a very straightforward concept of norms as simply static constraints for a single agent. In conclusion, the different architectures reveal different aspects of norms because they are concerned with different target systems.

Moreover, the control of the causal power of the respective components is dependent on the complexity of the system. A system analysis can yield more significant conclusions if the system is more simple. Thus, there is a trade-off between the accuracy of the description and its explanatory power.

Table 8 A hypothetical best choice

Backgr.	Soc. emb.	Phil. back.	Concepts	Dynamics	Conflicts
BD	Repres. Social	Consequ. Deontic	Concept	Innovation Spreading	Bet. agents Goal/norm Bet. norm

Acknowledgments This work has been undertaken as part of the Project 'Emergence in the Loop' (EmiL: IST-033841), funded by the Future and Emerging Technologies programme of the European Commission, within the framework of the initiative "Simulating Emergent Properties in Complex Systems". I would like to thank anonymous referees for helpful comments and advises and for a stimulating discussion at the World Congress on Social Simulation 2008.

References

1. Andrighetto G, Campennì M, Conte R, Paolucci M (2007) On the immergence of norms: a normative agent architecture. In: Proceedings of AAAI symposium, Social and organizational aspects of intelligence, Washington DC, pp 11–18
2. Axelrod R (1986) An evolutionary approach to norms. Am Polit Sci Rev 80:1095–1111
3. Boella G, Lesmo L (2001) Deliberate normative agents: basic instructions. In: Conte R, Dellarocas C (eds) Social order in multiagent systems. Kluwer, Norwell, pp 65–77
4. Boella G, van der Torre L (2003) Norm Governed Multiagent Systems: the delegation of control to autonomous agents. In: Proceedings of the IEEE/WIC IAT conference, IEEE Press, pp 10–27
5. Boella G, van der Torre L (2006) An architecture of a normative system: counts-as conditionals, obligations, and permissions. In: AAMAS, ACM, pp 229–231
6. Boman M (1999) Norms in artificial decision making. Artif Intell Law 7:17–35
7. Broersen J, Dastani M, Huang Z, van der Torre L (2001) The BOID architecture: conflicts between beliefs, obligations, intentions, and desires. In: Proceedings of the 5th international conference on autonomous agents, pp 9–16
8. Broersen J, Dastani M, van der Torre L (2001) Wishful thinking. In: Proceedings of DGNMR'01
9. Broersen J, Dastani M, van der Torre L (2005) Beliefs, obligations, intentions, and desires as components in an agent architecture. Int J Intell Syst 20:893–919
10. Castelfranchi C, Dignum F, Treur J (2000) Deliberative normative agents: principles and architecture. In: Jennings NR, Lesperance Y (eds) LNCS, vol 1757. Springer, Berlin, pp 364–378
11. Conte R, Castelfranchi C (1995) Cognitive and social action. UCL Press, London
12. Conte R, Castelfranchi C (1999) From conventions to prescriptions. Towards an integrated view of norms. Artif Intell Law 7:119–125
13. Conte R, Dellarocas C (2001) Social order in info societies: an old challenge for innovation. In: Conte R, Dellarocas C (eds) Social order in multiagent systems. Kluwer, Norwell, pp 1–16
14. Conte R, Dignum F (2001) From social monitoring to normative influence. J Artif Soc Soc Simul 4. http://www.soc.surrey.ac.uk/JASSS/4/2/7.html
15. Dignum F, Kinny D, Sonenberg L (2002) From desires, obligations and norms to goals. Cogn Sci Q 2. http://people.cs.uu.nl/dignum/papers/CSQ.pdf
16. Garcia-Camino A, Rodriguez-Aguilar JA, Sierra C, Vasconcelos W (2006) Norm-oriented programming of electronic institutions: a rule-based approach. In: AAMAS 2006, ACM, pp 33–40
17. Grimm V (2006) A standard protocol for describing individual-based and agent-based models. Ecol Modell 198:115–126
18. Hegselmann R (2009) Moral dynamics. In: Meyers R (ed) Encyclopedia of complexity and system science. Springer, Berlin
19. Kutschera F von (1999) Grundlagen der Ethik. deGruyter, Berlin
20. Lopez F, Marquez A (2004) An architecture for autonomous normative agents. In: 5th Mexican international conference in computer science, ENC 04, IEEE Computer Society, Los Alamitos, USA, pp 96–103
21. Lotzmann U, Möhring M (2009) Simulating norm formation – an operational approach. In: Proceedings of the 8th International Conferenc on autonomous agents and multi-agent systems (AAMAS 2009), pp 1323–1324

22. Neumann M (2008) Homo socionicus: a case study of simulation models of norms. J Artif Soc Soc Simul 11/4. http://jasss.soc.surrey.ac.uk/11/4/6.html
23. Ormerod P (2008) What can agents know? The feasibility of advanced cognition in social and economic systems. In: Proceedings of the AISB 2008 convention: communication, interaction and social intelligence, Aberdeen. http://www.aisb.org.uk/convention/aisb08/proc/proceedings/06%20Agent%20Emergence/03.pdf
24. Rao AS, Georgeff MP (1991) Modeling rational agents within a BDI architecture. In: Proceedings of the KR91, pp 473–484
25. Richiardi M, Leombruni R, Saam N, Sonnessa M (2006) A common protocol for agent-based social simulation. J Artif Soc Soc Simul 9. http://jasss.soc.surrey.ac.uk/9/1/15.html
26. Sadri F, Stathis K, Toni F (2006) Normative KGP Agents. Comput Math Organ Theory 12:101–126
27. Shoham Y, Tenneholtz M (1992) On the synthesis of useful social laws for artificial agent societies (preliminary report). In: Proceedings of the 10th AAAI conference, pp 276–281
28. Skyrms B (1996) Evolution of the social contract. Cambridge University Press, Cambridge
29. Skyrms B (2004) The stage hunt and the evolution of social structure. Cambridge University Press, Cambridge
30. Vazquez-Salceda J, Aldewereld H, Dignum F (2005) Norms in multiagent systems; from theory to practice. Int J Comput Syst Eng 20:225–236
31. Verhagen H (2001) Simulation of the learning of norms. Soc Sci Comput Rev 19:296–306
32. von Wright G (1950) Deontic logic. Mind 60:1–15
33. Young P (1998) Social norms and economic welfare. Eur Econ Rev 42:821–830
34. Young P (2003) The power of norms. In: Hammerstein P (ed) Genetic and cultural evolution of cooperation. MIT Press, Cambridge, MA, pp 398–399

The Complex Loop of Norm Emergence: A Simulation Model

Giulia Andrighetto, Marco Campennì, Federico Cecconi, and Rosaria Conte

Abstract In this paper the results of several agent-based simulations, aiming to test the effectiveness of norm recognition and the role of normative beliefs in norm emergence are presented and discussed. Rather than mere behavioral regularities, norms are here seen as behaviors spreading to the extent that and because the corresponding commands and beliefs do spread as well. More specifically, we will present simulations aimed to compare the behavior of a population of normative agents provided with a norm recognition module and a population of social conformers whose behavior is determined only by a rule of imitation. The results of these simulations show that under specific conditions, i.e. moving from one social setting to another, imitators are not able to converge in a *stable way* on one single behavior; vice-versa, normative agents (equipped with the norm recognition module) are able to converge on one single behavior.

Keywords Norm emergence • Norm immergence • Normative architecture • Norm recognition • Agent based social simulation

1 Introduction

Traditionally, the scientific domain of normative agent systems presents two main directions of research. The first, related to Normative Multiagent Systems, focuses on *intelligent* agent architecture, and in particular on normative agents and their capacity to *decide* on the grounds of norms and the associated incentive or sanction. This scientific area absorbed the results obtained in the formalization of normative

G.A ndrighetto (✉), M. Campennì, F. Cecconi, and R. Conte
LABSS, Istituto di Scienze e Tecnologie della Cognizione, CNR,
viaS.M artinode llaB attaglia44, 0018 5,R ome,I taly
e-mail:giulia .andrighetto@istc.cnr.it

K. Takadama et al. (eds.), *Simulating Interacting Agents and Social Phenomena: The Second World Congress*, Agent-Based Social Systems 7,
DOI 10.1007/978-4-431-99781-8_2, © Springer 2010

concepts, from deontic-logic [24], to the theory of normative position [15], to the dynamics of normative systems. Those studies have provided Normative Multiagent Systems with a formal analysis of norms, thus giving crucial insights to *represent* and *reason* upon norms. The second is focused on much simpler agents and on the *emergence of regularities* from agent societies.

Very often, the social scientific study of norms goes back to the philosophical tradition that defines norms as regularities emerging from reciprocal expectations [3, 11]. Indeed, interesting sociological works [17] point to norms as public goods, the provision of which is promoted by second-order cooperation [14]. This view inspired most recent work of evolutionary game-theorists [13], who explored the effect of *punishers* or *strong reciprocators* on the group's fitness, but did not account for the individual decision to follow a norm.

No apparent contamination and integration between these different directions of investigation has been achieved so far. In particular, it is unclear how something more than regularities can emerge in a population of intelligent autonomous agents and whether agents' *mental capacities* play any relevant role in the emergence of norms.

The aim of this paper is help to clarify what aspects of cognition are essential for norm emergence. We will concentrate on *one* of these aspects, i.e. *norm recognition*. We will simulate agents endowed with the ability to tell what a norm is, while observing their social environment.

One might question why start with norm recognition. After all, isn't it more important to understand *why* agents observe norms? Probably, it is. However, whereas this question has been answered to some extent [7] the question how agents tell norms has received poor attention so far.

In this paper, we will address the antecedent phenomenon, the *norm recognition*, postponing the consequent, the norm compliance, to future studies. In particular, we will endeavour to show the impact of norm recognition on the emergence of a norm. More precisely, we will observe agents endowed with the capacity to recognize a norm (or a behavior based on a norm), to generate new normative beliefs and to transmit them to other agents by communicative acts or by direct behaviors.

We intend to show whether a society of such normative agents allows norms to emerge. The notion of norms that we refer to [8] is rather general. Unlike a *moral* notion, which is based on the sense of right or wrong, norms are here meant in the broadest sense, as behaviors spreading to the extent that and because (a) they are prescribed by one agent to another, (b) and the corresponding normative beliefs spread among these agents (for a more detailed characterization of social norms see Sect. 3).

Again, one might ask why not to address our moral sense, our sense of the right or wrong. The reason is at least twofold. First, our norms are more general than moral virtues. They include also social and legal norms. Secondly, and moreover, agents can deal with norms even when they have no moral sense: they can even obede norms they believe to be injust. But in any case, they must know what a norm is.

2 Existent Approaches

Usually, in the formal social scientific field, and more specifically in utility and (evolutionary) game theory [11, 20], the spread of new norms and other cooperative behaviors is not explained in terms of internal representations. The object of investigation is usually the conditions for agents to converge on given behaviors, which proved efficient in solving problems of coordination or cooperation, independently of the agents' normative beliefs and goals [4]. In this field, no theory of norms based on mental representations (of norms) has yet been provided.

Game theorists essentially aimed at investigating the dynamics involved in the problem of norm convergence. They considered norms as conditioned preferences, i.e. options for actions preferred as long as those actions are believed to be preferred by others as well [3]. Here, the main role is played by *sanctions*: what distinguishes a norm from other cultural products like values or habits is the fact that adherence to a social norm is enforced by many phenomena, including sanctions [2, 12]. The utility function, which an agent seeks to maximize, usually includes the cost of sanction as a crucial component (see Sect. 3 for an alternative view of norms).

In the field of multi-agent systems [23], on the contrary, norms are explicitly represented. However, they are implemented as *built-in* mental objects. This alternative approach has been focused on the question as to how autonomous intelligent agents decide on the grounds of their explicitly represented norms.

Even when norm emergence is addressed [19], the starting point is some preexisting norms, and emergence lies in integrating them. When agents (with different norms) coming from different societies interact with each other, their individual societal norms might change, merging in a way that might result to be beneficial to the societies involved (and the norm convergence results in the improvement of the average performance of the societies under study).

Lately, decision making in normative systems and the relation between desires and obligations has been studied within the Belief-Desire-Intention (BDI) framework, developing an interesting variant of it, i.e. the so-called Belief-Obligations-Intentions-Desires or BOID architecture [5]. This is a feedback loops mechanism, which considers all the effects of actions before committing to them, and resolves conflicts among the outputs of its four components. Examples for such approach are [16]. Obligations are introduced to constrain individual intentions and desires on the one hand, while preserving individual autonomy on the other. Agents are able to violate normative obligations. This implies that agents dispose of the skill of normative reasoning.

In evolutionary psychology, there are many efforts to define normative agents. Each definition focuses on one specific aspect of the problem. To make one good example, in Sripada and Stichs model [22], mechanisms of norm acquisition are proposed, but no description how they work is given. In none of these approaches, including the last one, it is possible for an agent to tell that a given input is a (new) norm. On the contrary, obligations are *hard-wired* into the agents' minds when the system is off-line. Unlike the game-theoretic model, multi-agent systems certainly exhibit all of the advantages deriving from an explicit representation of norms. Nevertheless, we claim that it suffers from some limitations, which have not only a theoretical, but also a practical and implementative relevance.

First of all, multi-agent systems overshadow one of the advantages of autonomous agents, i.e. their capacity to filter external requests. Such a filtering capacity affects not only normative decisions, but also the acquisition of new norms. Indeed, agents take decisions even when they decide to form normative beliefs, and then new (normative) goals, and not only when they decide whether to execute the norm or not [9].

As to the practical relevance, if agents are enabled to acquire new norms, there is no need for exceedingly expanding their knowledge-base, since they may be optimized when they are on-line [21].

Despite the undeniable significance of the results achieved, these studies leave some fundamental questions still unanswered, such as how and where norms originate and how agents acquire norms. In sum, our feeling is that the question how norms are created and innovated has not received so far the attention it deserves. We claim that this circumstance may be ascribed the way the normative agent has been modeled up to now.

The aim of this paper is to propose a computational model of autonomous norm recognition and to test its effectiveness in norm emergence through a set of agent based simulations. We claim that a capacity for autonomous norm-recognition would greatly enhance Multiagent Systems' flexibility and dynamic potentials. The ability to generate normative beliefs from external (observed or communicated) inputs makes an autonomous social agent more adaptable to the social settings that she encounters and at the same time more proactive in the spreading of norms.

3 Social Norms

As already said, in many economic and sociological approaches to norms, there is a strong, if not exclusive, emphasis on sanctions as necessary for the existence of norms. In such a view, norms are aimed to create and impose constraints on the agents' in order to obtain a given coordinated collective behaviour. The other side of norms is often ignored: their function of inducing new goals into the agents' minds in order to influence them (not) to do something. Norms do not operate just limiting some possible choices. Norms also work by adding new alternatives and generating new goals: they influence us to do something that might have never entered in our mind otherwise.

In order to fulfill the abovementioned functions, norms require a distinctive nature. We consider a social norm as a social behavior that spreads trough a population thanks to the diffusion of a particular shared belief, i.e. the *normative-belief*. A normative belief, in turn, is a belief that a given behavior, in a given context, for a given set of agents, is either forbidden, obligatory, permitted, etc. Behaviours may be perceived as prescribed without having being explicitly issued [8]. It has to be pointed out that the prescription through which a norm is transmitted is a special one: a prescription that is requested to be adopted because it is a norm and it is fully applied only when it is complied with for its own sake. In order for the norm to be satisfied, it is not sufficient that the prescribed action is performed, but it is necessary to comply with the norm *because of* the normative goal, that is, the goal deriving from the recognition and subsequent adoption of the norm.

Thus, for a norm-based behavior to take place, a normative belief has to be generated into the minds of the norm addressees, and the corresponding normative goal has to be formed and pursued. Our claim is that a norm emerges as a norm only when it is incorporated into the minds of the agents involved [7, 8]; in other words, when agents *recognize* it as such. In this sense, norm emergence and stabilization implies its *immergence* [6] into the agents' minds.

This meaning that norm emergence is a two way dynamics, consisting of two processes: (a) emergence: process by means of which a norm not deliberately issued spreads through a society; (b) *immergence*: process by means of which a normative belief is formed into the agents minds [1, 6, 10]. Emergence of social norms is due to the agents' behaviors, but the agents' behaviors are due the mental mechanisms controlling and (re)producing them (immergence). Adopting this view of norms is not for nothing. It requires that we shed light on how norms work through the minds of the agents, and how they are represented.

4 Objectives

Aiming to test the effectiveness of the norm recognition module, we run several agent-based simulations in which very simple agents interact in a common hypothetical world. This very abstract model serve to the purpose to check (a) whether there are crucial differences at the population level, between Social Conformers (SCs), whose behavior is determined only by imitation, and Norm Recognizers (NRs), whose behavior is instead affected by a norm recognition module and (b) whether and to what extent the capacity to recognize a norm and to generate normative beliefs is a preliminary requisite for norm emergence.

The structure of the paper is as follows. After a short description of our normative agent architecture (EMIL-A) (see [1]), we focus on the norm recognition module (Sect. 6.1). In Sect. 7, we present a simulation model aimed to test the effectiveness of the norm recognizer. In Sect. 8 the simulator is described. Finally, in Sect. 9 results are discussed.

5 Finding Out Norms

Norms are social artifacts, emerging, evolving, and decaying. If it is relatively clear how legal norms are put into existence, it is much less obvious how the same process may concern social norms. How do new social norms and conventions come into existence?

Some simulation studies about the selection of conventions have been carried out, for example Epstein and colleagues' study of the emergence of social norms [11], and Sen and Airiau's study of the emergence of a precedence rule in the traffic [20]. However, such studies investigate which one is chosen from a set of alternative equilibria. A rather different sort of question concerns the emergence and innovation of social norms when no alternative equilibria are available for selection. This subject is still not widely investigated and references are scanty, if any (see [18]).

We propose that a possible answer to the puzzling questions posed above might be found out examining the interplay of communicated and observed behaviors, and the way they are represented into the minds of the observers. If any new behavior α is interpreted as obeying a norm (for a detailed description of the norm recognition process see par. 5.1), a new *normative belief* is generated into the agent's mind and a process of normative *influence* will be activated. Such a behavior will be more likely to be replicated than in the case if no normative belief had been formed [1]. As shown elsewhere [1], when a normative believer replicates α, she will influence others to do the same not only by ostensibly exhibiting the behavior in question, but also by explicitly conveying a norm. People impose new norms on one another by means of *deontics* and explicit *normative valuations* (for a description see par. 5.1) and propose new norms (implicitly) by means of (normative) behaviors. Of course, having formed a normative belief is necessary but not sufficient for normative influence: we will not answer the question *why agents do* so (a problem that we solve for the moment in probabilistic terms), but we address the question how they can influence others to obey the norms. They can do so if they have formed the corresponding normative belief, if they know how one ought to behave.

Hence we propose that normative recognition represents a crucial requirement of norm emergence and innovation, as processes resulting from both agents' interpretations of one another's behaviors, and their transmitting such interpretations to one another.

6 Normative Architecture

6.1 Norm Recognizer

Our normative architecture (EMIL-A) (see [1] for a detailed description) consists of mechanisms and mental representations allowing norms to affect the behaviors of autonomous intelligent agents. EMIL-A is meant to show that norms not only regulate the behavior but also act on different aspects of the mind: recognition, adoption, planning, and decision-making. Unlike BOID in which obligations are already implemented into the agents' minds, EMIL-A is provided with a component by means of which agents infer that a certain norm is in force even when it is not already stored in their normative memory. Implementing such a capacity is conditioned to modeling agents' ability to recognize an observed or communicated social input as normative (Fig. 1), and consequently to form a new normative belief. In this paper, we will only describe the first component of EMIL-A, i.e. the norm recognition module. This is most frequently involved in answering the open question we have raised, i.e. how a new norm is found out. Topic that we consider particularly crucial in norm emergence, innovation and stabilization.

Our Norm Recognizer (see Fig. 2) consists of a long term memory, the normative board, and in a working memory, presented as a three layers architecture. The normative board contains normative beliefs, ordered by *salience*. With salience we refer to the degree of activation of a norm: in any particular situation,

Fig. 1 The social input

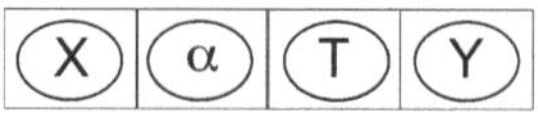

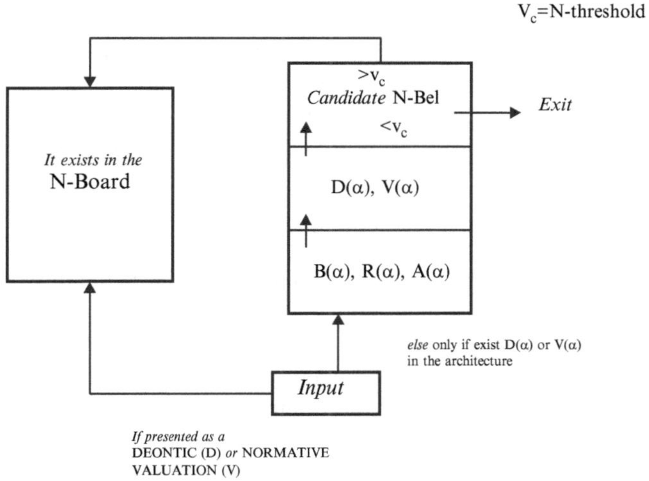

Fig. 2 The norm recognition module (in action): it includes a long term memory (*on the left*), i.e. the normative board, and a working memory (*on the right*). The working memory is a three layer architecture, where the received input is elaborated. *Vertical arrows* in the block *on the right side* indicate the process regulating the generation of a new normative belief. Any time a new input arrives, if contains a deontic (D), for example "you must answer, when asked" or a normative valuation (V), for example "Not answering when asked is impolite", a candidate belief "One must answer, when asked" is generated and temporally stored at the third level of the working memory. If the normative threshold is exceeded, the candidate normative belief will leave the working memory and stored in the normative board as a normative belief. After the candidate normative belief "One must answer, when asked" has been generated, the normative threshold can be reached in several ways: one way consists in observing for a given number of time agents performing the same action (α) prescribed by the candidate normative belief, i.e. agents answering when asked. If the agent receives no other occurrences of the input action (α), after a fixed time t, the candidate normative belief will leave the working memory (see Exit)

one norm may be more frequently followed than others, its salience being higher. The difference in salience between normative beliefs and normative goals has the effect that some of these normative mental objects will be more active than others and they will interfere more frequently and with more strength with the general cognitive processes of the agent.[1]

The working memory is a three layer architecture, where social inputs are elaborated. Agents observe or communicate *social inputs*. Each input (see Fig. 1) is presented as an ordered vector, consisting of four elements:

- The source (X), i.e. the agent from which we observe or receive the input
- The action transmitted (α), i.e. the potential norm
- The type of input (T): it can consist either in a *behaviour* (B), i.e. an action or reaction of an agent with regard to another agent or to the environment, or in a *communicated* message, transmitted through the following holders:
  - Assertions (A), i.e. generic sentences pointing to or describing states of the world
  - Requests (R), i.e. requests of action made by another agent
  - Deontics (D), partitioning situations between good (acceptable) and bad (unacceptable). Deontics are holders for the three modal verbs analyzed by von Wright [24]: "may" (indicating a permission), "must" (indicating an obligation) and "must not" (indicating a prohibition)
  - Normative valuations (V), i.e. assertions about what it is right or wrong, correct or incorrect, appropriate or inappropriate (i.e. *it is correct to respect the queue*)
- The observer (Y), i.e. the observer/addressee of the input

The input we have modelled is far from accounting for the extraordinary complexity of norms. In fact, a variety of items can operate as inputs, allowing new candidate beliefs to be generated into the agents' minds. Several tests should be executed in order to evaluate whether the received input is actually a norm. For example, the *source* should be carefully evaluated, examining whether she is entitled to issue the norm. This test is supported by other more specific investigations relative to some features of the source; for example: Is the norm within the source's domain of competence? Is the current context the proper context of the norm? Is the addressee within the scope of the source's competence? The *motives* for issuing a norm should be carefully evaluated too. More specifically, it has to be checked if the norm has been issued for personal or private interest, rather than for the interest the source is held to protect. If the norm addressee believes that the prescription is only

[1] At the moment, the normative beliefs' salience can only increase, depending on how many instances of the same normative belief are stored in the Normative Board. This feature has the negative effect that some norms become highly salient, exerting an excessive interference with the decisional process of the agent. We are now improving the model, adding the possibility that, if the normative belief is inactive for a certain amount of time, its salience will decrease.

due to some private desire, she will not consider it as a norm. Only after this evaluation process (that we have not implemented yet), a normative belief can be generated.

Once received the input, the agent will compute, thanks to its norm recognition module, the information in order to generate/update her normative beliefs. Here follows a brief description on how this normative module works. Every time a message containing a deontic (D), for example, "You must answer when asked", or a normative valuation (V), for example "Not answering when asked is impolite", is received, it will directly access at the second layer of the architecture, giving rise to a candidate normative belief "One must answer when asked", which will be temporally stored at the third layer. This will sharpen agents' attention: further messages with the same content, especially when observed as open behaviors, or transmitted by assertions (A), for example "When asked, Paul answers", or requests (R), for example "Could you answer when asked?", will be processed and stored at the first level of the architecture. Beyond a certain normative threshold (which represents the frequency of corresponding normative behaviors observed, e.g. n% of the population), the candidate normative belief will be transformed in a new (real) normative belief, that will be stored in the normative board, otherwise, after a certain period of time, it will be discarded.

Aiming to decide which action to produce, the agent will search through the normative board: if more than one item is found out, the most salient norm will be chosen.

7 The Simulation Model

In order to test the effectiveness of norm recognition on norm emergence, and to see what are the observable effects of norm recognition, we have implemented two different kinds of agents, Social Conformers (SCs) (see Sect. 7.2) and Norm Recognizers (NRs) (see Sect. 7.3), and we have compared their behaviours in an artificial and stylized environment.

In our simulation model the environment consists of four scenarios, in which the agents can produce three different kinds of actions. We define two context-specific actions for every scenario, and one action common to all scenarios. Therefore, we have nine actions. To see why, suppose that the first context is a post office, the second an information desk, the third our private apartment, and so on. In the first context *stand in the queue* is a context-specific action, whereas in the second a specific action could be *occupy a correct place in front of the desk*. A common action for all of the contexts could be *answer when asked*.

SCs and NRs, each of them is provided with a personal agenda (i.e. a sequence of contexts), an individual and constant time of permanence in each scenario (when the time of permanence is expired, the agent moves to the next context) and a window of observation (i.e. a capacity for observing and interacting with a fixed number of agents) of the actions produced by other agents.

In addition, NRs are also provided with the norm recognition module.

7.1 Moving Across Scenarios

The agents can move across scenarios: once expired the time of permanence in one scenario, each agent – be she a SC or a NR – moves to the subsequent scenario following her agenda. Such an irregular flow (each agent has a different time of permanence and a different agenda) generates a complex behavior of the system, tick-after-tick producing a fuzzy definition of the scenarios, and tick-for-tick a fuzzy behavioral dynamics.

7.2 Social Conformers

At each tick, two SCs are paired randomly and allowed to interact. The action that each agent produces is affected by the actions performed by previous n agents (*social conformity rate*): if there is an action that has been carried out more often than the others, the agent will imitate that action. Otherwise, she will randomly choose one among the three possible actions for the current scenario. For the aim of this work, we have opted for a *fitness-independent heuristics*; we are not interested in the individual performance of the agents, but in detecting the observable effects of different ways of modeling the emergence of norms. We run several simulations, with different social conformity rates' values.

7.3 Norm-Recognizers

At each tick, unlike SCs, the NRs (paired randomly) interact exchanging social inputs.

NRs produce different behaviors: if the normative board of an agent is empty (i.e. it contains no norms), the agent produces an action randomly chosen from the set of possible actions (for the context in question); in this case, also the type (T) of input by means of which the action is communicated is chosen randomly. Vice versa, if the normative board contains some norms, the agent chooses the action corresponding to the most salient among these norms. In this case, the agent can either communicate the norm by a D or a V, or perform an action compliant with it (we are improving the model giving the agent the possibility to violate the norm). This corresponds to the intuition that if an agent has a normative belief, there is a high propensity (in this paper, this has been fixed to 90%) for her to transmit it to other agents under normative modals (D or V) or a complaint behavior (B).

We ran several simulations varying the number of agents and the value of the threshold that causes the generation of new normative beliefs.

8 The Simulator

We realized a simulator for non-expert users with a simple interface, such that the user can manipulate the number of agents involved in the simulation, the number of time's ticks, the number of contexts and the population type, whether composed by SCs or by NRs. Simulations with mixed populations are not allowed at this stage. Other variables that the user can modify include the social conformity rate and standard routines, e.g., save the results of simulation in a workspace. The simulation is implemented by means of Matlab functions.

9 Results and Discussion

We briefly summarize the simulation scheme. As already said, we call SCs the agents performing imitation and NRs the agents equipped with a norm-recognizer. We compared a population of SCs with a population of NRs, modifying the parameters concerning the strength of the imitative process in a way that will be clarified in detail ahead.

Both in the case of SCs and NRs, the process begins by producing actions (and modals in the case of NRs) at *random*, and continues with agents conforming themselves to others.

The assimilation process is synchronic. During the simulation, the agent in position i provides inputs to the agents in i-1, i-2, i-k positions, and results are assigned immediately. If the number of agents before i does not reach the threshold, i.e. if the agent does not observe enough agents performing the same action, no imitation takes place. Imitation among SCs is implemented by means of a *voting mechanism*, such that agent *i* performs the *most frequent action* among those that have been executed by the agents who preceded him.

In the NRs case, the process is more complex: agent *i* provides inputs to the agent who precedes her (k=1). Action choice is conditioned by the state of her normative board. When all of the agents have executed one simulation update, the whole process restarts at the next step.

9.1 Results with Social Conformers

In Fig. 3 we show the distribution of the performed actions accomplished by 100 SCs during 100 simulation ticks. On axis x the time flows from time $t=0$ to time $t=100$ corresponding to the end of the simulation. On axis y the number of performed actions for each different type of action is indicated.

The results shown in Fig. 4 are very clear: SCs do not produce social norms. In fact, we can see that any convergence towards one single action appears through the time: the convergence rate does not change significantly. The *convergence rate*

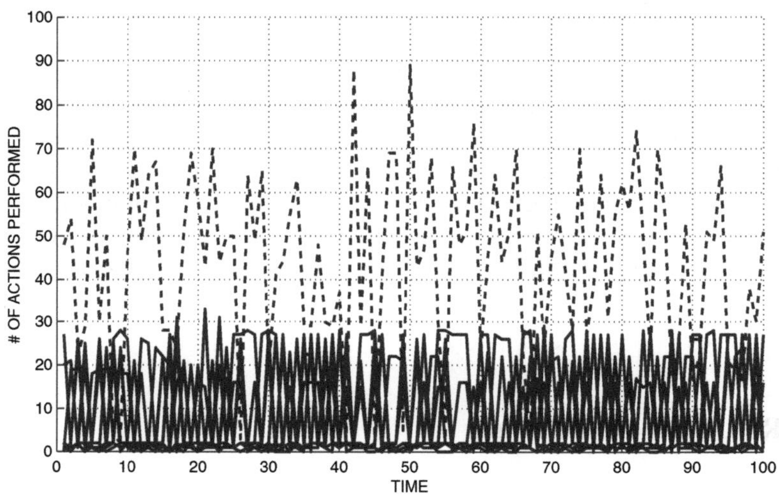

Fig. 3 Actions performed by SCs. On axis X the number of simulation ticks (100) are indicated and on axis Y the number of performed actions for each different type of action. The *dash line* corresponds to the action common to all scenarios

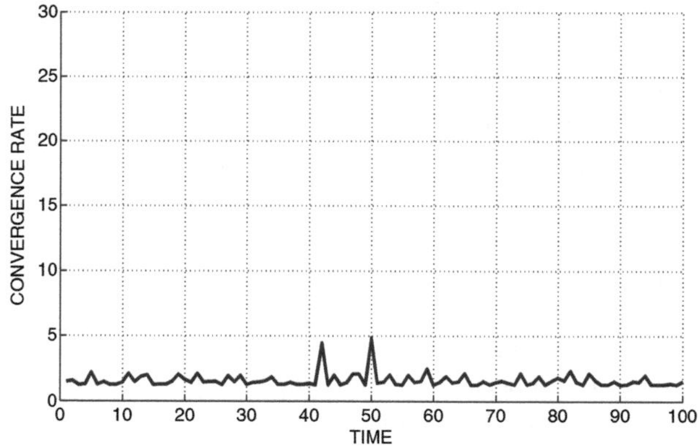

Fig. 4 On axis X, the flow of time is indicated; on axis Y the value of convergence rate

indicates the agents' convergence on single actions: more is the convergence on a specific action, more is the value of this rate. In Fig. 3, for $t=100$, the distribution of actions at tick=100 is shown. The common action (dash line) is the most frequent. We could call action 1 an *imitation norm*, the only feature of which is *frequency*.

9.2 Results with Norm Recognizers

The situation looks rather different among NRs. Figures 5 and 7 show actions distributions and normative beliefs for a certain value of the norm threshold: in Fig. 5 we show the actions distribution, in Fig. 7 we show the overall number of normative beliefs generated through the time into the NRs' normative boards. To be noted, a normative belief is not necessarily the most frequent belief in the population. However, norms are behaviors that spread thanks to the spreading of the corresponding normative belief. Figure 6 shows that in the NRs population we can appreciate a strong convergence towards a single action; despite SCs (where convergence rate is stable through the time), the NRs' convergence rate *increases* through the time (compare the convergence rate in SCs' a NRs' populations).

Hencefore, we can summarize some results for NRs as follows: normative beliefs lead to (a) a clear convergence towards an action and to (b) a stronger variance between different actions (compare for example Figs. 5 and 3). In Fig. 5, after tick=60, we can appreciate a significant growth in the number of one performed action, related to the common action (dash line), due to the effect of normative beliefs acting on the choice of agents' behaviors. In other words, after

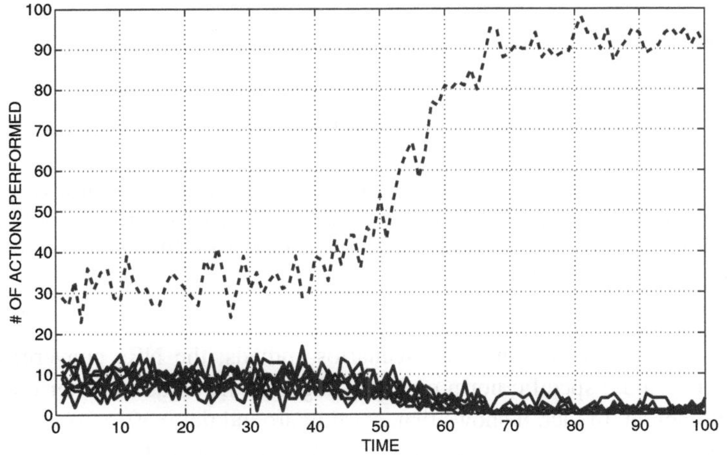

Fig. 5 Actions performed by NRs. On axis X, the number of simulation ticks (100) is indicated and on axis Y the number of performed actions for each different type of action. The *dash line* corresponds to the action common to all scenarios

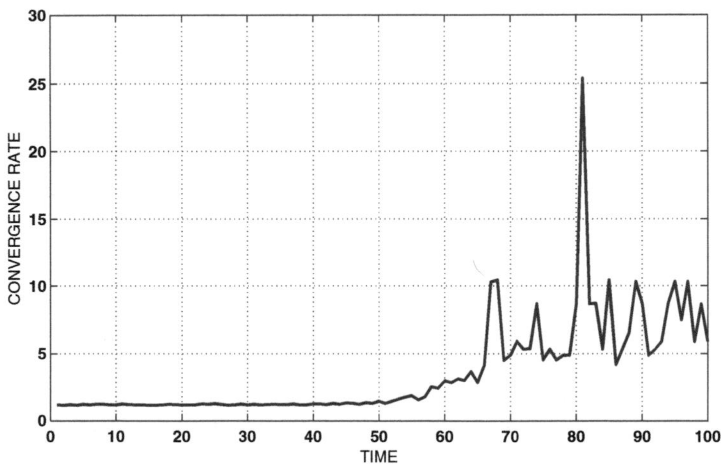

Fig. 6 On axis X, the flow of time is shown; on axis Y the value of convergence rate

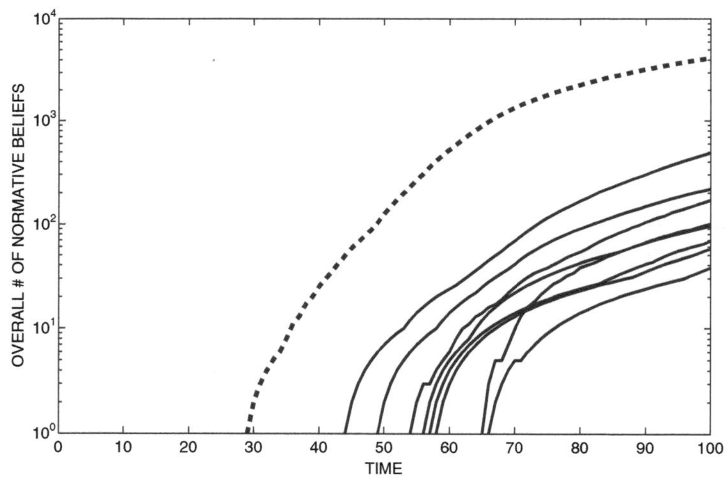

Fig. 7 Each line corresponds to the trend of the number of new normative beliefs' generation through the time. On axis X we have simulation's ticks; on axis Y the number of new normative beliefs. This figure shows the results with 100 agents and 100 ticks. The *dash line* corresponds to the action common to all scenarios

tick=60 the action common to all scenarios spreads; the NRs converge on one single action and a specific norm emerges. Figure 7 helps us to better understand this phenomenon. In fact, it shows that, starting around tick=30, a normative belief (the one related to the common action 1: dash line) appears in the agents' normative boards and starts to increase, in spite of the spreading of action 1 that become uniform after tick=60 (see Fig. 5). There is a time interval of 30 ticks between the normative belief appearance (tick=30) and the agents convergence on the common action 1 (tick=60). It is interesting to observe that during this interval other normative beliefs are generated and stored into the agents' minds, although the earlier is more frequent.

For a normative belief to affect the behavior, a certain number of ticks has to elapse, which we might call *norm latency*.

Recognizing the existence of this time interval between the N-belief appearance and the convergence on the corresponding action has an important impact on the theory of norms we have presented in Sect. 3. The immergence process occurs before the emergence one, but it takes time for effect to occur. Norm emergence is a two way dynamics consisting of an immergence process followed by an emergence one, generating a virtuous loop.

The NRs' population has a *pool* of potential social norms (see Fig. 7), corresponding to poorly salient normative beliefs, just because variance in each context and in each tick is strong.

10 Concluding Remarks

The findings presented above seem to suggest that the way of modeling the two populations of agents reach convergence in fairly different ways: Social Conformers (SCs) tend, within the same simulation tick, to converge in a *homogenous* way, due to the fact that behaviour is strongly influenced by their neighbours. They converge *en masse* on one single action, the *imitation norm*, rapidly but *unstably*: conformity varies over time, depending on the actions of the other agents within each scenario and over time.

Conversely, Norm Recognizers (NRs) converge while preserving their *autonomy*: they choose how to act considering the normative beliefs they have formed while observing and interacting with others.

Thus, they converge in a stable way: after a certain period of time the majority of agents start to perform the same action. It is possible to say that a norm has emerged after having immerged into the agents' minds.

In sum, norm recognition seems to represent a crucial requirement of norm emergence, as the process resulting from both agents' interpretations of one anothers behaviours, and of their transmitting such interpretations to one another.

In fact, the ability to generate normative beliefs with different degrees of salience from external (observed or communicated) inputs gives more stability to the emerged process.

Follow-ups of this work will introduce improvements, regarding both social and cognitive refinements. In fact, at the moment, on one hand, only the norm-recognition module is actually driving the agents' behaviors, while the other procedures, such as the Norm Adoption, the Decision Making and the Normative Action Planning, are still inactive. On the other hand, normative social dynamics, such as punishment or enforcement mechanisms, have not been implemented yet. In future works, it will be interesting to design experiments (a) mixing SCs and

NRs, (b) mixing NRs with different normative thresholds to see what difference this would make.

We consider our simulative platform a theoretical tool allowing us to test the potentiality and limitations of the present norm recognition module, and indicating the necessity for augmenting it in order to model further interesting intra and inter-agent dynamics.

Acknowledgments This work was supported by the EMIL project (IST-033841), funded by the Future and Emerging Technologies program of the European Commission, in the framework of the initiative Simulating Emergent Properties in Complex Systems.

References

1. Andrighetto G, Campennì M, Conte R, Paolucci M (2007) On the immergence of norms: a normative agent architecture. In: Proceedings of AAAI symposium, social and organizational aspects of intelligence, Washington, DC
2. Axelrod R (1986) An evolutionary approach to norms. Am Polit Sci Rev 4(80):1095–1111
3. Bicchieri C (2006) The grammar of society: the nature and dynamics of social norms. Cambridge University Press, New York
4. Binmore K (1994) Game-theory and social contract, vol 1. Fair playing. Clarendon, Cambridge
5. Broersen J, Dastani M, Hulstijn J, Huang Z, van der Torre L (2001) The BOID architecture. Conflicts between beliefs, obligations, intentions and desires. In: Proceedings of the fifth international conference on autonomous agents, Montreal, Quebec, Canada, pp 9–16
6. Castelfranchi C (1998) Simulating with cognitive agents: the importance of cognitive emergence. In: Multi-agent systems and agent-based simulation, Heidelberg
7. Conte R, Castelfranchi C (1995) Cognitive and social action. University College of London Press, London
8. Conte R, Castelfranchi C (2006) The mental path of norms. Ratio Juris 19(4):501–517
9. Conte R, Castelfranchi C, Dignum F (1998) Autonomous norm-acceptance. In: Proceedings of the 5th international workshop on intelligent agents V, agent theories, architectures, and languages, pp 99–112
10. Conte R, Andrighetto G, Campenni M, Paolucci M (2007) Emergent and immergent effects in complex social systems. In: Proceedings of AAAI symposium, social and organizational aspects of intelligence, Washington, DC
11. Epstein J (2006) Generative social science. Studies in agent-based computational modeling. Princeton University Press, Princeton, New Jersey
12. Feld T (2006) Collective social dynamics and social norms. Munich Oersonal RePEc Archive
13. Gintis H, Bowles S, Boyd R, Fehr E (2003) Explaining altruistic behavior in humans. Evol Hum Behav 24:153–172
14. Horne C (2007) Explaining norm enforcement. Rationality and Society 19(2):139–170
15. Lindhal L (1977) Position and change. Reidel Publishing Company, Dordrecht
16. Lopez y Lopez F, Luck M, d'Inverno M (2002) Constraining autonomy constraining autonomy constraining autonomy through norms. In: AAMAS '02
17. Oliver PE (1993) Formal models of collective action. Annu Rev Sociol 19:271–300
18. Posner RA, Rasmusen EB (1999) Creating and enforcing norms, with special reference to sanctions. Int Rev Law Econ 19(3):369–382
19. Savarimuthu B, Purvis M, Cranefield S, Purvis M (2007) How do norms emerge in multi-agent societies? Mechanisms design. The Information Science Discussion Paper (1)

20. Sen S, Airiau S (2007) Emergence of norms through social learning. In: Proceedings of the twentieth international joint conference on artificial intelligence
21. Shoham Y, Tennenholtz M (1992) On the synthesis of useful social laws in artificial societies. In: Proceedings of the 10th national conference on artificial intelligence, Kaufmann. San Mateo, California, pp 276–282
22. Sripada C, Stich S (2006) The innate mind: culture and cognition. Oxford University Press, Oxford, Chap A Framework for the Psychology of Norms
23. Van der Torre L, Tan Y (1999) Contrary-to-duty reasoning with preference-based dyadic obligations. Ann Math Artif Intell 27(1–4):49–78
24. von Wright GH (1963) Norm and action. A logical inquiry. Routledge and Kegan Paul, London

A Social Network Model of Direct Versus Indirect Reciprocity in a Corrections-Based Therapeutic Community

Nathan Doogan, Keith Warren, Danielle Hiance, and Jessica Linley

Abstract Therapeutic communities (TCs) for substance abuse depend heavily on mutual aid between residents, who are expected to help each other in attaining recovery. An example of this mutual aid is that residents are expected to affirm each other for actions that are considered helpful in recovery or beneficial to the community. It is unclear how the residents maintain cooperative behavior. In this paper we construct an agent-based model in which affirmations are recorded as directional arcs in a social network. Reception of affirmations can be based on either direct reciprocity, an exchange of aid between two agents, friend of a friend dynamics, in which the exchange extends by one node, or indirect reciprocity, in which helping one agent improves the reputation of the agent who gives help and thereby attracts help from others. We then compare the model results to records of affirmations between residents kept in a TC. We find that indirect reciprocity more closely mimics the level of reciprocity and transitivity found in the actual network.

Keywords Therapeutic community • Agent-based model • Social network • Cooperation • Game theory

1 Introduction

Residential therapeutic communities (TCs) are the most common form of substance abuse treatment in the American correctional system. Such TCs house as many as 150 residents at one time. TCs are based on mutual aid between residents, who are expected to support each other in recovering from substance abuse. TC researchers have gone so far as to say that the community of fellow substance abusers is the method of treatment [5]. Qualitative studies have suggested that TC residents do

N. Doogan (✉), K. Warren, D. Hiance, and J. Linley
College of Social Work, The Ohio State University,
1947 College Road, Columbus, OH 43210, USA.
e-mail: doogan.1@osu.edu; warren.193@osu.edu; hiance.1@osu.edu; linley.8@osu.edu

K. Takadama et al. (eds.), *Simulating Interacting Agents and Social Phenomena: The Second World Congress*, Agent-Based Social Systems 7,
DOI 10.1007/978-4-431-99781-8_3, © Springer 2010

value the help of their peers, but take several months to internalize the TC culture of mutual aid [4]. At this point there has been no quantitative study of the maintenance of cooperation in a TC.

TCs present a challenge to process researchers because of the large number of residents, the importance of mutual aid between them, and the consequent necessity of understanding how residents maintain cooperation through interpersonal interactions that occur dozens or even hundreds of times per day. Agent-based models are a powerful and flexible methodology for analyzing interactions between individuals [7, 12]. They are good candidates to exceed the limitations of previous process research methods.

In this paper we develop a simple model of interpersonal interaction in a TC, based on known mechanisms of direct and indirect reciprocity [1, 13]. The behavior modeled is the TC practice of affirming fellow residents for actions that help others and themselves advance toward recovery [5, 10]. Affirmations are meant both to encourage residents and reward them for actions that further the goals of the community and the recovery of their peers [5]. Affirmations are public, both because the resident who is affirmed knows about them and because they are often announced during some public time, for instance during meetings or during meals [10]. The model saves each affirmation as a directed arc in a social network [3, 19]. We compare the resulting network with empirical social network data derived from clinical records kept at a corrections-based TC.

2 Theoretical Background

2.1 Direct and Indirect Reciprocity

Since evolutionary and game theoretic models of human behavior assume self-interest, both have difficulty in explaining mutual aid between individuals who are notge neticallyre lated[13].T rivers[17]ha spropos eda mode lofdire ctre ciprocity in which individuals interact repeatedly and can either cooperate or defect on any round. Cooperation can therefore evolve when the likelihood of a second meeting with a partner exceeds the cost-benefit ratio of defection versus cooperation [13]. In a famous series of computer tournaments, Axelrod and Hamilton [2] demonstrated that a strategy known as tit-for-tat, in which an agent begins by cooperating with a partner and thereafter does whatever the partner did in a previous round, was capable of leading to cooperative behavior in groups.

Experiments have demonstrated that human beings often practice direct reciprocity regardless of whether it maximizes their utility. For instance, Falk [8] sent letters soliciting a donation to roughly 10,000 participants. Each participant was randomly assigned either to receive no gift, a small gift of one post card or a larger gift of four postcards. Both frequency and size of donations increased with the size of the gift. In a meta-analysis of methods aimed at improving the return rate of surveys, Edwards et al. [6] found that response rate nearly doubled when an incentive that was not conditioned on response – i.e., a gift – was included with the questionnaire.

Indirect reciprocity is a term used for two distinctly different phenomena. In the first, sometimes known as "downstream indirect reciprocity" [13, 15], an individual who helps another acquires a reputation for helping; this reputation, in turn, attracts aid from third parties. Thus, an individual does not need to actually meet the beneficiary of a helping act again in order to receive a benefit [13, 15]. Downstream indirect reciprocity promotes cooperation if the probability of knowing an individual's reputation exceeds the cost to benefit ratio of a cooperative act. Nowak [13] has compared direct reciprocity to barter and downstream indirect reciprocity to the invention of money.

Experiments indicate that humans often practice downstream indirect reciprocity. In experimental conditions, participants who donate to UNICEF receive more money from co-participants than those who do not [11]. While some studies indicate that third parties who give to their more generous helpers themselves receive added benefit from fourth parties [20], downstream indirect reciprocity occurs even in the absence of actual material benefits [11, 21]. Individuals will help someone who is helping others even if they do not personally stand to gain.

The second form of indirect reciprocity is the tendency of individuals to help others after they themselves have received help. Sometimes known as upstream indirect reciprocity, it has been observed in experiments with human beings [9, 12] and also with rats [16]. While upstream indirect reciprocity has been repeatedly observed in experiments, it has been difficult to construct a theoretical explanation, because it confers no benefit on the reciprocator. However, Nowak and Roch [14] have pointed out that upstream indirect reciprocity can lead to random walks of cooperative acts that jump from one member of the group to another, adding to the overall level of cooperation in the group. Further, upstream indirect reciprocity can evolve as a side effect of direct reciprocity.

2.2 *Indirect and Direct Reciprocity in TCs*

Affirmations in TCs are made in public, and when recorded are actually read aloud during meal time so as to positively reinforce the actions of those being affirmed [10]. While affirmations are intended to help the recipient, their public nature means that there is the automatic side effect that anyone who gives one will be identified as someone who helps others. TC clinical theory suggests that peers should regard such an individual as a role model [5]. Downstream indirect reciprocity will occur if role models attract further affirmations because of their reputation. TC clinical literature also suggests that affirmations can increase residents' motivation to help others [5], an example of upstream indirect reciprocity.

On the other hand, direct reciprocity in the form of giving an affirmation to another resident simply to return a previous affirmation is not encouraged. This is because the point of an affirmation is to acknowledge prosocial behavior and encourage the recipient [5]. It is not clear that simply affirming those who have previously affirmed oneself accomplishes these goals.

It therefore seems likely that a simple model based on downstream indirect reciprocity will generate a social network that closely reproduces the network of affirmations that actually occurs in a TC. On the other hand, it seems unlikely that a model based on direct reciprocity would perform as well.

3 Methodology

3.1 Agent-Based Model

The agent-based model was created using NetLogo version 4.0.2 [22]. The model simulates individual agents' admissions into the TC, length of stay, interactions, and graduations from the TC. The model allows the user to modify the way local interactions occur between agents during their residence in the TC.

The key user-changeable settings in the model are used to modify the methods by which the affirmation sender and the affirmation receiver are selected and the environment in which these agents operate. While the model is running, it keeps track of clock ticks. The number of clock ticks in 1 day equals the number of participants (90 for the runs reported in this paper) multiplied by two. For each tick of the clock the model must determine if an affirmation will be sent. This decision is based on the setting that controls the ratio of ticks at which an affirmation will be sent, thus allowing the user to adjust the mean number of affirmations sent per day.

Once the decision is made that some agent will send an affirmation at a particular tick, two subroutines are enacted to choose the sender and receiver – in that order – of the affirmation. The sender subroutine chooses an agent through a lottery in which a variable called confidence determines the weights. When an agent receives an affirmation, it gains two units of confidence, and is therefore more likely to send an affirmation in return at a later tick. This models upstream indirect reciprocity. Confidence declines at a rate set by the user. This rate could vary from zero – confidence could never decline – to two – confidence declines by two units after 1 day. It is possible to set the daily decline higher than two, but this eventually leads to a high level of instability in confidence as the daily decline becomes as large as or larger than any possible daily increase. For the runs reported in this paper we set the rate of confidence decline to 0.4, allowing for a slow decline. However, confidence only declines if its value is higher than the taper rate. Thus, confidence can never go below zero and once it has attained a level equal to the taper rate it cannot go below that level. It is therefore assumed that receiving affirmations has at least a small enduring effect, an assumption that is supported by the literature on TCs [18].

Once the model has decided on an appropriate sender for a particular tick of time, it will then choose a receiver for the affirmation based on user choice of one of three subroutines. The first subroutine models direct reciprocity, and also involves a lottery. However, in this lottery, tickets are only given to the subgroup of agents who have previously affirmed the sender. Figure 1 illustrates the direct reciprocity model.

In the second subroutine, tickets are given to agents who have previously affirmed the sender and also a second layer of agents who have affirmed the agents

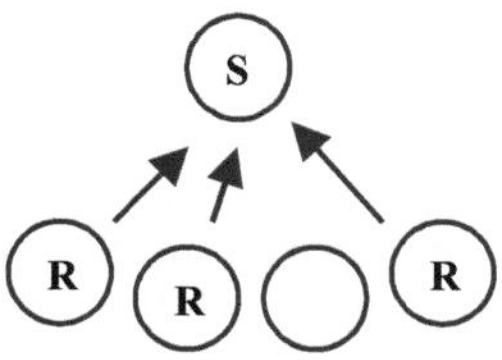

Fig. 1 Diagram illustrating the direct reciprocity model. The receiver is selected from a group of potential receivers (R) who have affirmed the sender (S) in the past. The *arrows* represent affirmations sent at a prior time point. The *circle with no link to the current sender* represents an agent who has not affirmed the sender, and who will therefore not be considered by the sender as a potential receiver. The agent thus lacks an "R" label

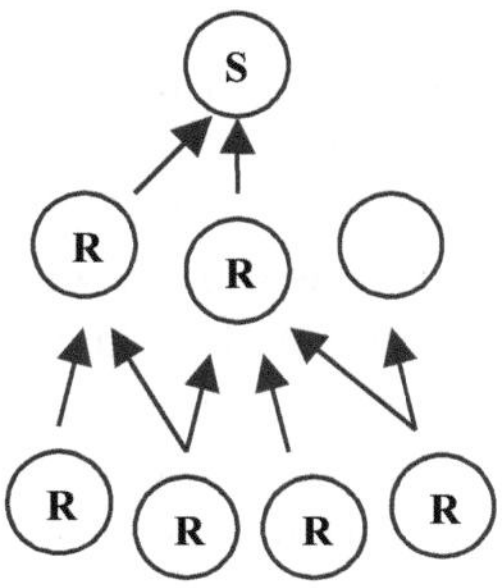

Fig. 2 Diagram illustrating the friend of a friend model. This is the same as the direct reciprocity model except the sender "S" also considers the group of agents that have affirmed his/her "friends" as possible receivers. (A "friend" is simply someone who has affirmed the sender.) Again, the agent with no label is not considered because it has no link to "S." The "friend" of the non-labeled agent is considered by "S" because it also has a second link to a "friend" of "S"

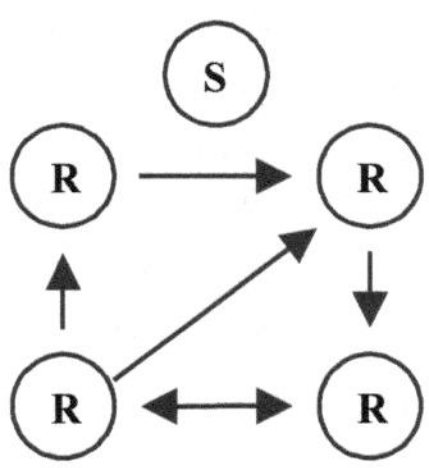

Fig. 3 Diagram illustrating the indirect reciprocity model. In this model, the sender "S" considers all agents regardless of their connection with "S," based on whether they have affirmed one or more other agents. In this diagram, no agent is connected with "S." But because they have all sent at least one affirmation to another agent, "S" considers each to be a possible receiver

who affirmed the sender. We refer to this as the friend of a friend model. Because it seems nonsensical that an individual would reciprocate at one person removed without also reciprocating directly, we run this model in conjunction with the model of direct reciprocation. Figure 2 illustrates the friend of a friend model.

The third model uses downstream indirect reciprocity; the receiver of the affirmation is selected by a lottery where a particular agent gets a number of lottery tickets equal to the number of affirmations the agent has sent. (This mechanism is a modified version of the Netlogo 4.02 model of preferential network attachment. Rather than nodes which have incoming links receiving more, those that have outgoing links receive more.) Figure 3 illustrates the indirect reciprocity model.

These are parsimonious models that do not include strategic behavior on the part of the agents. However, it is not clear that mapping out joint payoff functions is the most relevant way of applying game theory to clinical situations such as therapeutic communities. Moreover, experimental evidence has repeatedly indicated that, in human beings at least, direct reciprocity, upstream indirect reciprocity and downstream indirect reciprocity all act in the absence of utility maximization on the part of one or more of the interacting individuals [8, 11, 21]. Given these considerations, a parsimonious model that does not attempt to specify strategic behavior is justified.

3.2 Data

The data for this analysis are drawn from one Ohio therapeutic community, West Central Community Correctional Facility (WCCCF), which keeps extensive records of resident affirmations and corrections. These records are kept in order to monitor program fidelity to the best standards of TC practice, the overall functioning of the community and the progress of individual residents. In recording an affirmation, residents fill out a form that includes the individual giving the affirmation, the individual receiving it, the content and the date. Once the form is filled out, a committee of senior residents and staff members vet the affirmation for legitimacy. For instance, if one resident affirms another for praising drug use, the affirmation will be disallowed. Once the committee has declared the affirmation to be legitimate, it is announced to the entire community – usually during lunch – and is recorded in a computer database. In total, WCCCF recorded 65,952 affirmations from 1,031 residents to other residents between June 6, 2001 and January 31, 2006, the period covered by this analysis. This does not, of course, represent all of the affirmations that residents gave during that period; numerous informal, unrecorded affirmations occur as well. However, the recorded affirmations are particularly important to residents because they are publicly announced and entered into a permanent record.

All WCCCF residents in this data set are male, and approximately 85% of them are of European-American origin. Age at the time the data was collected ranged from 18 to 60. Mean length of stay was 146 days and median length of stay was 163 days, with a maximum of 195 days and a minimum of 1 day. Fewer than 1% of all residents stayed beyond the theoretical maximum of 180 days, and fewer than 20% stayed for less than 4 months.

3.3 Analysis

Both empirical and model data was analyzed using the UCINET 6.0 social network analysis program. WCCCF records of affirmations were combined into an edgelist format for the sake of analysis, with each row in the list including sender, receiver, and number of affirmations sent. In social network terms, each row included a sender, a receiver and a valued arc. The simulation output was also combined into an edgelist.

Because the affirmations in our data set were publicly announced, we expected that simulation using indirect reciprocity would most closely reproduce the empirical network. We chose six network statistics for comparison. The first two were Freeman degree centrality measures of the variability of indegrees and outdegrees between residents. The measures compare the centrality of a given network with a maximally central network in which all indegrees or outdegrees focus on one node. A lower value indicates a less centralized network [19, pg. 199]. We chose these measures of centrality because previous research indicates that both giving and receiving affirmations predicts success following graduation [18]. One should therefore expect a fairly decentralized network of affirmations in an effective TC. The second two measures looked at reciprocity; if A affirms B, is B also likely to affirm A? These measure whether residents are exchanging affirmations. We would expect them to be quite different depending on whether residents were directly exchanging affirmations or giving them to peers who had established a reputation for helping others. The former case should lead to a higher measure of reciprocity than the latter. Reciprocity can be measured in two different ways: the percentage of dyads that are reciprocal and the percentage of arcs that are part of reciprocal dyads. We report both. The fifth measure was transitivity. If A affirms B, and B affirms C, then A should also affirm C [19, pgs. 243–247]. Transitivity is measured as the percentage of ordered triples in which $i \to j$ and $j \to k$ in which $i \to k$. Transitivity is a measure of the ability of two individuals to reach agreement on affirming a third. It should be higher if indirect reciprocity is an important determinant of affirmations, because residents should be able to reach agreement on who among their peers most deserves to be affirmed. Finally, we included the percentage of triangles with at least two arcs that have a third, a broader measure of triangle formation.

All of these measure characteristics of participants' local networks, albeit averaged over the entire network of interactions. We did not choose measures that look at agents' relationship to the entire network, such as betweenness centrality or closeness centrality, for two reasons. First, there is empirical evidence that local interactions influence outcomes; those individuals who give and receive more affirmations, i.e. have more indegrees and outdegrees, are less likely to be reincarcerated [18]. Second, global centrality measures were largely developed to understand power relations, on the assumption that those individuals who are more central are also more powerful [19]. It is difficult to see how this reasoning applies to the network of affirmations that is the object of this study.

4 Results

All models were run with a total of 90 agents on any given day. Each agent was active in the model for 154 days, the midpoint between the mean and median residences in the program. The agent then graduated and was immediately replaced by another agent. The model was run for the same number of days as were in the data set, and the mean number of affirmations per day was the same as in the data set. Results are in Table 1.

Both the models and the empirical data yield highly decentralized networks. Direct reciprocity appears to more closely mimic the degree centrality of the data set than either indirect reciprocity or the friend of a friend model, coming within roughly 1% of the true value of outdegree centrality. All modeled networks are even closer to the observed values of indegree centrality. However, all the networks are so decentralized that no really large differences appear. The indirect reciprocity

Table 1 Measures of centrality, reciprocity and transitivity under three rules of exchange and in one empirical dataset. All measures are expressed in percentages

Measure	Direct reciprocity (%)	Friend/friend (%)	Indirect reciprocity (%)	WCCCF (%)
Out-degree entrality	0.785	0.227	0.443	1.58
In-degree centrality	0.424	0.305	0.419	0.972
Dyad based reciprocity	62.64	70.36	39.68	34.87
Arc based reciprocity	77.03	82.60	56.82	51.71
Transitivity*	14.29	11.01	41.79	39.88
Triangle completion**	5.16	3.95	19.28	16.84

*Percentage of ordered triples in which i→j and j→k that are transitive
**Percentage of triangles with at least two legs that have a third

model, on the other hand, fairly closely tracks the reciprocity and transitivity values of the empirical data set. For indirect reciprocity, 39.68% of all dyads are reciprocal, versus 34.87% in the empirical network and 56.82% of all arcs are part of reciprocal dyads, versus 51.71% of all arcs in the empirical dataset. The indirect reciprocity model also tracks the transitivity results closely; 41.79% of all ordered triples that could be transitive are, while the same is true of 39.88% in the empirical data. In the indirect reciprocity model, 19.24% of all triangles with two arcs have a third; in the empirical data set the percentage is 16.84%. Both the direct reciprocity and friend of a friend model yield much higher levels of reciprocity and much lower levels of transitivity than the empirical dataset. While these are descriptive statistics, the difference between models is obvious.

These values make it clear that the model of direct reciprocity creates an inherent tension between reciprocity and the creation of triads. Under direct reciprocity, positive feedback means that directed arcs between dyads become more common, thus reducing the number of arcs available to form triads. This reduces the likelihood that transitive triads will form and the likelihood that any triad that happens to have two arcs will have a third. The friend of a friend model does not change this tendency, because it says when i→j and j→k, then k→i. This is not a transitive relationship. Also, if i→j and j→k, k is still likely to reciprocate directly to j rather than indirectly to i.

The tension does not disappear if we change the rate at which the confidence variable, with which we model upstream indirect reciprocity, changes. We have analyzed networks modeled with both direct reciprocity and friend of a friend reciprocity over a broad range of rates at which confidence declines. We have been unable to tune either of these models in such a way that it reproduces either the reciprocity or transitivity that is found in the empirical data set. For instance, if confidence never declines in the friend of a friend model, dyad-based reciprocity is 76.55%, arc-based reciprocity is 86.72%, 9.67% of ordered triples in which i→j and j→k are transitive, and 3.44% of all triangles with at least two arcs have a third. On the other hand, if confidence declines by two units per day, dyad-based reciprocity is 63.75%, arc-based reciprocity is 77.87%, 6.34% of ordered triples in which i→j and j→k are transitive, and 2.14% of all triangles with at least two arcs have a third. Results with the model of direct reciprocity have been similar.

5 Conclusion

This study is clearly exploratory in nature; data is gathered only at one site, analysis is entirely through descriptive social network statistics, and we only matched one run of each model to the data. Future research will include replication at other sites, a sufficient number of runs to establish confidence intervals around the theoretical parameters, and more emphasis on network time series analysis. However, we do believe that this exploratory work carries important implications for both research and practice in TCs.

Of the three agent-based models in this paper, that based on downstream indirect reciprocity does the best job of reproducing observed patterns in the empirical data. This supports the hypothesis that downstream indirect reciprocity plays an important role in TCs. It is not obvious from a reading of TC literature that clinical researchers clearly understand this. Announcement of affirmations is meant to encourage and positively reinforce those who receive them. The focus of the affirmation is clearly on the recipient [6]. However, it is inevitable that a public announcement will attract attention to the individual who gave the affirmation. The close match between the downstream indirect reciprocity model and the empirical data suggests that this can lead to indirect reciprocity.

In light of the TC emphasis on role models, it is striking that the model of indirect reciprocity closely reproduced the level of transitivity found in the data. A role model is, by definition, someone whom others have agreed to respect. In a transitive relationship defined by the exchange of affirmations, i shows his respect by affirming j, j shows his respect by affirming k, and i, who has already affirmed j, affirms k as well. Thus, i and j agree on their respect for k, and transitivity can be seen as a local measure of the production of role models. The relatively high level of transitivity in the indirect reciprocity model, when compared to the other two, suggests that indirect reciprocity at least does not impede the production of a local network of role models, and may facilitate the emergence of such a network.

On the other hand, both the model of direct reciprocity and the friend of a friend model do a very poor job of reproducing network characteristics. They both produce far higher levels of reciprocity and far lower levels of transitivity than can be found in the empirical data. These models bind the agents in reciprocal dyads, in which an affirmation from A to B makes it more likely that B's next affirmation will go to A, thereby making it more likely that A's next affirmation will go to B, and so on. This is a very powerful positive feedback loop. Because these models produce such a high level of reciprocity, they simply use up the affirmations that could lead to more numerous transitive relationships. These models imply that reciprocity in affirmations cannot offer a complete explanation of the structure of this network.

TC clinicians often display concern about whether residents are simply reciprocally exchanging affirmations. Such exchange is viewed as less valuable for a resident's recovery than a broader practice of affirming all those who have earned an affirmation (Carole Harvey, personal communication, February 12, 2008). The social network analysis in this article suggests that this concern may additionally be justified on the grounds that direct reciprocity impedes the development of social structure beyond reciprocal dyads, and in particular impedes the development of transitive triads.

Finally, this analysis shows that an approach to process research in residential programs and other group based interventions that combines agent based models and social network analysis is feasible and potentially fruitful. Traditional quantitative process research in such programs, like other quantitative research in the social sciences, has largely ignored the interactions between individuals. Qualitative researchers, drawing on ethnographic and grounded theory traditions, have addressed such interactions, but have thus far shown little interest in developing more formal models. Agent-based models are an ideal methodology for generating hypotheses on the interaction of large numbers of residents, and social network analysis offers a natural way of testing those hypotheses. Their combination promises to profoundly alter our understanding of residential therapeutic communities.

References

1. Axelrod R (1984) The evolution of cooperation. Basic Books, New York
2. Axelrod R, Hamilton WD (1981) The evolution of cooperation. Science 211(4489): 1390–1396
3. Barabasi AL (2002) Linked: the new science of networks. Perseus, Cambridge, MA
4. Burnett K (2001) Self-help or sink-or-swim? The experience of residents in a UK concept-based therapeutic community. In: Rawlings B, Yates R (eds) Therapeutic communities for the treatment of drug users. Jessica Kingsley, London, England
5. De Leon G (2000) The therapeutic community: theory, model and method. Springer, New York
6. Edwards P, Roberts I, Clarke M, DiGuiseppe C, Pratap S, Wentz R, Kwan I (2002) Increasing response rates to postal questionnaires: a systematic review. Br Med J 324(1183):1–9
7. Epstein J (2007) Generative social science: studies in agent-based computational modeling. Princeton University Press, Princeton, NJ
8. Falk A (2007) Gift exchange in the field. Econometrica 65(5):1501–1511
9. Fischbacher U, Gächter S, Fehr E (2001) Are people conditionally cooperative? Evidence from a public goods experiment. Econ Lett 71:397–404
10. Harvey C (ed) (2005) Why we do what we do: rationale and theoretical applications for therapeutic community activities in Ohio. Ohio Department of Alcohol and Drug Addiction Services, Columbus, OH
11. Milinski M, Semmann D, Krambeck H (2002) Donors to charity gain in both indirect reciprocity and political reputation. Proc R Soc Lond B 269:881–883
12. Miller JH, Page SE (2007) Complex adaptive systems: an introduction to computational models of social life. Princeton University Press, Princeton, NJ
13. Nowak MA (2006) Five rules for the evolution of cooperation. Science 314:1560–1563
14. Nowak MA, Roch S (2007) Upstream reciprocity and the evolution of gratitude. Proc R Soc B Biol Sci 274(1610):605–609

15. Nowak MA, Sigmund K (2005) Evolution of indirect reciprocity. Nature 437:1291–1298
16. Rutte C, Taborsky M (2007) Generalized reciprocity in rats. PLoS Biol 5(7):16–21
17. Trivers RL (1971) The evolution of reciprocal altruism. Q Rev Biol 46:35–57
18. Warren K, Harvey C, DeLeon G, Gregoire T (2007) I am my brother's keeper: affirmations and corrective reminders as predictors of reincarceration following graduation from a corrections-based therapeutic community. Offender Substance Abuse Report 7(3):33–43
19. Wasserman S, Faust K (1994) Social network analysis: methods & applications. Cambridge University Press, Cambridge
20. Wedekind C, Braithwaite VA (2002) The long-term benefits of human generosity in indirect reciprocity. Curr Biol 12:1012–1015
21. Wedekind C, Milinski M (2000) Cooperation through image scoring in humans. Science 288:850–852
22. Wilensky U (1999) NetLogo. Center for connected learning and computer-based modeling. Northwestern University, Evanston, IL. http://ccl.northwestern.edu/netlogo

A Force-Directed Layout for Community Detection with Automatic Clusterization

Patrick J. McSweeney, Kishan Mehrotra, and Jae C. Oh

Abstract We present a force-directed layout algorithm to detect community structure within a network. Our algorithm places nodes in a two-dimensional grid and continuously updates their positions based on two opposing forces: nodes pull connected nodes closer and push non-connected nodes away. To compute the strength of the forces between nodes we make use of two insightful community properties: high-degree nodes contribute more inter-community edges than low-degree nodes and the graph-distance between two nodes is inversely proportional to the probability that they belong to the same community. We present empirical evidence in support of both of these claims. In conjunction, we use a clustering algorithm to monitor and interpret the current community structure. Running our algorithm on well-known social networks, we find that we produce accurate results that avoid some common pitfalls of alternative approaches.

Keywords Social network analysis • Community detection

1 Introduction

The community structure of a complex network characterizes the decomposition of topological information into meaningful modules. These modules are regions in a network that have a higher ratio of intra-module edges than inter-module edges. Community detection is the problem of discovering these regions in a given network. This research problem has attracted attention from researchers from the fields of biology [11], physics [19] and social science [21].

In order to devise an algorithm for detecting community structure, we first must formally define what we mean by community. The starting point for all definitions of community is the intuition that a community is a set of nodes which have a

P.J. McSweeney (✉), K. Mehrotra, and J.C. Oh
SyracuseU niversity, Syracuse,N Y,U SA
e-mail: pjmcswee@syr.edu; mehrotra@syr.edu; jcoh@syr.edu

K. Takadama et al. (eds.), *Simulating Interacting Agents and Social Phenomena: The Second World Congress*, Agent-Based Social Systems 7,
DOI 10.1007/978-4-431-99781-8_4, © Springer 2010

higher density of connections to each other than to the rest of the network. Let $G = <N, E>$ be a graph (complex network) and let $C \subseteq N$ be a subset of nodes (module). A community structure is a partition P of N. Elements C_i of the partition $P = \{C_1, .., C_r\}$ are the communities provided P covers the entire set of nodes $\bigcup_i^r C_i = N$ and for each $i \neq j$, C_i and C_j are disjoint[1] (i.e. $C_i \cap C_j = \emptyset$).

For a node $v \in C$ let $k_v^{in}(C) = |\{(v, v') \in E \wedge v' \in C\}|$ and let $k_v^{out}(C) = |\{(v, v') \in E \wedge v' \notin C\}|$. $k_v^{in}(C)$ is the number of edges that connect v to other nodes in C, and $k_v^{out}(C)$ is the number of other nodes that are connected to v but not in C. Radicchi [18] formalized definitions of strong community and weak community:

Strong Community: C is a strong community iff:

$$\forall v \in C(k_v^{in}(C) > k_v^{out}(C))$$

Weak Community: C is a weak community iff:

$$\sum_{v \in C} k_v^{in}(C) > \sum_{v \in C} k_v^{out}(C)$$

A strong community, C_s, is a set of nodes in which every node in C_s has more edges into C_s then out of C_s. A weak community, C_w, is a set of nodes in which there are more total edges that connect two nodes in C_w than there are total edges that connect a node in C_w to a node not in C_w.

We find Radicchi's definitions helpful but they do not directly give us a way to measure the degree to which a set of nodes exhibits the property of community structure. *How strong* is a given strong community structure, or for that matter a single community? To answer this question we turn to the popular modularity metric $Q(G,P)$ from Newman and Girvan [12, 13] for a given network *G* and a community structure *P* Let:

$$d(C) = \sum_{v \in C} degree(v)$$

$$l(C) = |\{(v, v') \in E \wedge v, v' \in C\}|.$$

Then the Newman and Girvan modularity measure is:

$$Q(G,P) = \sum_{C_i \in P} \left(\frac{l(C_i)}{|E|} - \left(\frac{d(C_i)}{2|E|} \right)^2 \right)$$

Q cumulatively measures the difference between the observed frequency of intra-community edges and the expected at-chance frequency of intra-community edges over all communities in the partition [12].

There are several existing algorithms that use local optimization techniques on Q: extremal optimization [3] and genetic algorithms [20] are just two of these algorithms. Q has become the standard for research into community detection.

---

[1]There are alternative interpretations of community that allow overlapping communities.

In this paper we discuss a new algorithm for detecting community structure that uses forces to form clusters of nodes, which can be placed in the family of other force-directed algorithms such as the Kamada-Kawai [10] graph layout algorithm.

We first discuss a specific weakness that we are trying to overcome. In Sect. 3 we discuss related research. Section 4 discusses two insights into community structure and presents empirical evidence to support these observations. Section 5 presents our algorithm and demonstrates it on a simple network. In Sect. 6, we present the results of our algorithm on several common social networks, and a special network that was designed [6] to illuminate shortcomings of other approaches. Section 7 discusses some improvements to automatically search out parameter values. Finally, we conclude and discuss our goals for future work.

2 A Weakness of Q

Let C_1, C_2 be two communities such that $d(C_1) = d(C_2) = d$. Let $Q_{C_i} = \frac{l(C_i)}{|E|} - \left(\frac{d(C_i)}{2|E|}\right)^2$ be the contribution made by C_i to Q, $i = 1,2$. Let $Q_{C_1 \cup C_2}$ be the corresponding contribution when C_1 and C_2 are joined together to form a single community. Fortunato and Barthelemy [6] have shown that:

$$Q_{C_1} + Q_{C_2} < Q_{C_1 \cup C_2} \text{ iff } \frac{d^2}{2|E|} < \#\text{ of edges between } C_1 \text{ and } C_2.$$

This result illustrates a drawback of Q for networks when $\frac{d^2}{2|E|} < 1$, i.e. when $d < \sqrt{2|E|}$, Q will prefer to merge the two communities even when visual (common sense) inspection suggests otherwise. Figure 1 displays an example in which this resolution limit occurs. In light of this revelation, we propose an algorithm which does not rely on Q for any of its computational strength.

Given the above shortcoming, we argue that currently there is no ideal, formalized definition of community structure. However, we resign ourselves to reporting our solutions using Q. This allows us to compare our results against existing algorithms, as Q is the predominately reported result.

3 Related Work

There are already several existing algorithms [2, 4, 7, 10, 16] which use forces to position nodes in a two or three dimensional grid, referred to as force-directed algorithms. The goal of the vast majority of these algorithms is to create an aesthetically appealing layout of the network, which may visually reveal certain network properties. As a point of interest we differ from the majority of these algorithms, in that we are not interested in any aesthetic value of our layout, although there may be some. In particular, we are only interested in detecting communities within these networks. LinLog [15, 16] is an existing algorithm with which we share the common goal of determining community structure through a force-directed

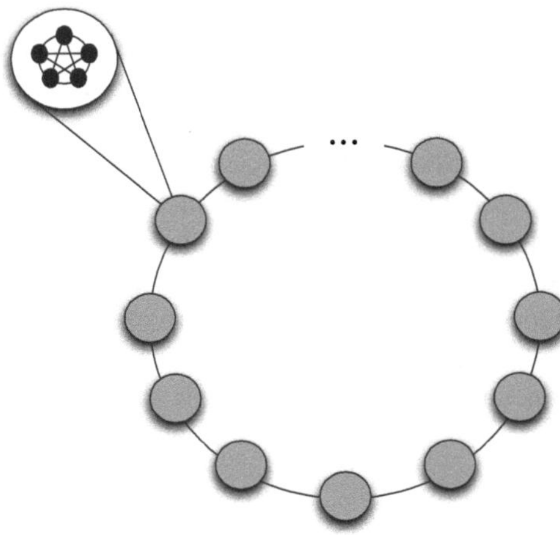

Fig. 1 Ring Network: Consider a network made up of 30 complete sub-graphs each with five nodes. Each subgraph is connected to its two neighboring complete sub-graphs by a single edge, forming a ring. A human observer will most likely select each of the 30 complete subgraphs as the community structure. However, notice that $d = 22$ and $\sqrt{2 \cdot |E|} \approx 35$, in particular $d < \sqrt{2|E|}$. Therefore maximization of Q will prefer to pair communities rather than individual communities themselves

algorithm. LinLog attempts to minimize the difference of position between adjacent nodes minus the natural log of all pairs of nodes.

$$U_{LinLog}(G = <N, E>) = \sum_{(u,v)\in E} \| \rho(u) - \rho(v) \| - \sum_{(u,v)\in N^2} ln(\| \rho(u) - \rho(v) \|),$$

where $\rho(u)$ is the position of node u and $\| \rho(u) - \rho(v) \|$ is the euclidean distance between node u and node v. The first term acts to attract together adjacent nodes and the second term acts to repel all pairs of nodes. There is also a slightly modified version of LinLog termed LinLog edge repulsion:

$$U^{edge}_{LinLog}(G =< N, E >) = \sum_{(u,v)\in E} \| \rho(u) - \rho(v) \| - \sum_{(u,v)\in N^2} degree(u) \times degree(v) \times ln(\| \rho(u) - \rho(v) \|),$$

which models repulsion between edges rather than nodes.

In addition to the difference in forces, which shall become clear after the next section, we also differ from LinLog in the significant way that our goal is to search out communities in networks, without explicitly optimizing a given utility function.

4 Community Structure Properties

Our algorithm relies on two simple but important observations made about the properties of community structures:

Observation i) In a scale-free network the degree of a node u is directly proportional to the number of inter-community edges that connect to u.

Observation ii) The graph distance, γ, (which is the minimum path length between two nodes) is inversely proportional to the probability that the two nodes are in the same community.

Both observations are not meant to be fixed laws but rather generalities that are true for many networks. To empirically examine these claims over a network we need to know the community structure of the network; however, we claim that these results are true for most reasonable community partition interpretations.

We have measured our observations in four well-known, undirected social networks: Zachary Karate Club [21], Dolphin Interaction network [11], Football Game Network [8] and Jazz Musician network [9]. We refer readers to the references for more details on these networks. We have used Q maximization for determining the community structure for the results that follow.

4.1 Node Degree

We measure *Observation i)* through the expression:

$$R(k) = \left(\frac{1}{|\psi(k)|}\right) \times \sum_{u \in \psi(k)} \frac{|(u,v) \in E \wedge v \notin C_u|}{2 \times \Phi_P},$$

where $\Psi(k)$ is the subset of nodes with degree k, Φ_P is the number of inter-community edges dictated by the community structure P and C_u is the community in P which contains u. $R(K)$ is the average percentage of inter-community edges that were contributed by nodes with degree k.

Figures 2a–d show the results for the social networks. The values have been normalized to have a maximum value at 1. The contribution to inter-community edges seems to increase with the degree of the node. The football network does not seem to follow this trend; our hypothesis is that the football network is not a scale-free network and this property is intricately linked to the scale-free property.

4.2 Graph Distance

We measure *Observation ii)* through the expression:

$$P(k) = \frac{|\{(u,v) : (\gamma(u,v) = k) \wedge (v \notin C_u)\}|}{|\{(u,v) : \gamma(u,v) = k\}|},$$

where $\gamma(u,v)$ is the graph distance between u and v, C_u is the community that contains u and k is a positive integer. The expression measures the proportion of

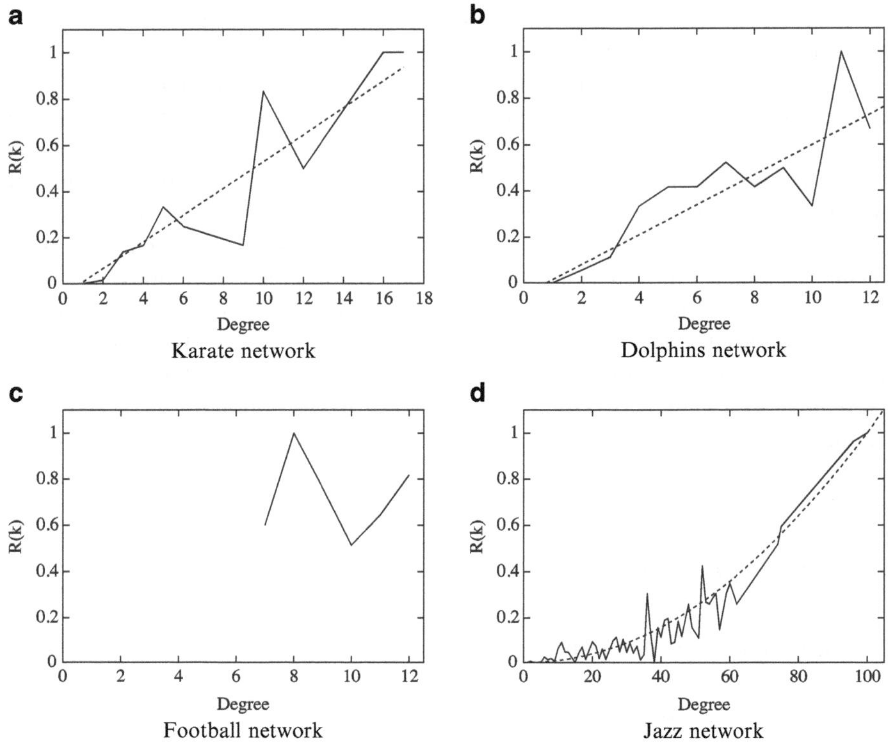

Fig. 2 The *dashed line* (when given) shows a fitted line for viewing. The *solid line* presents the empirical data collected. The x-axis for each image is the degree, the y-axis is the value of $R(K)$

nodes that are not in the same community and have graph-distance k, over the total number of node pairs that have graph-distance k.

Figures 3a–d show the results for the social networks. In these graphs, as the graph distance between two nodes increases the probability that the two nodes belong to different communities increases. It seems clear that most networks with community structure will show a similar relationship.

Based on the empirical data, we claim the probability of two nodes (graph distance k apart), given the diameter for the network $\delta(G)$, not being in the same community is:

$$\left(\frac{k}{\delta(G)}\right)^{\left(\frac{1}{k}\right)}$$

This is the dashed curve drawn in Fig. 3a–d.

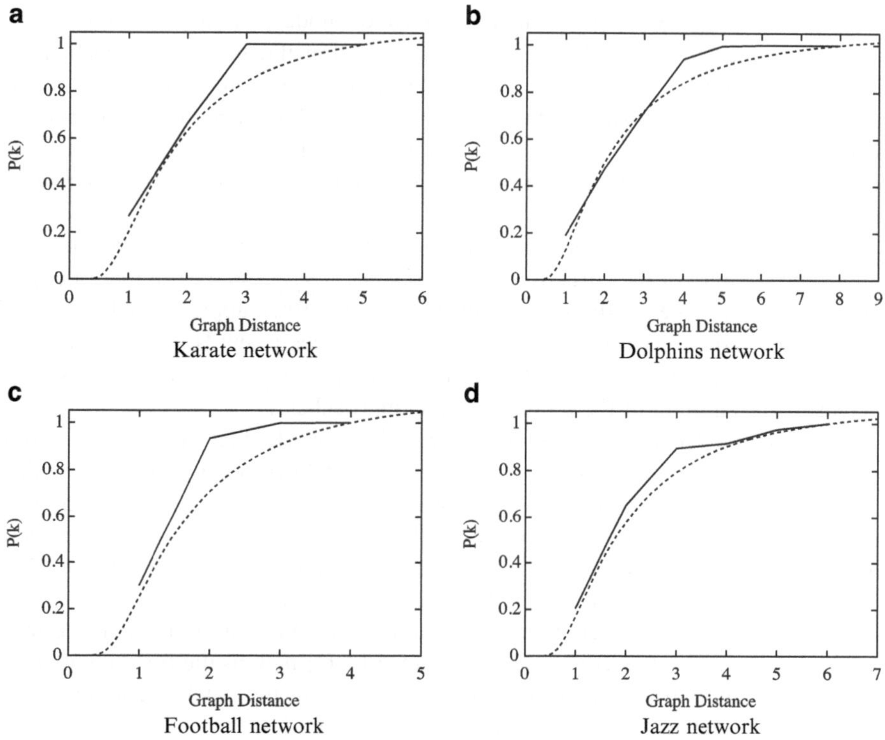

Fig. 3 The *dashed line* presents our estimation for the probability. The *solid line* presents the empirical data collected. The x-axis for each image is the graph-distance, the y-axis is the probability that two nodes are in the same community

5 Algorithms

In this section we present a theoretical framework and an algorithm based upon it in an effort to remove the use of utility functions in the context of searching for community structure. We circumvent the use of any explicit utility function and conceptually exploit Radicchi's definition of strong community, in which each node has more edges within its own community then with other communities.

There are two layers to our simulation: the force-directed layer and clustering layer. The force-directed layer runs a loop that continuously updates the position of every node. The clustering layer runs on top of the force-directed layer, its job is to detect clusters of nodes which represents the current community structure.

5.1 Force-Directed Layout

Each node is placed in a two-dimensional grid. The initial position for each node is a randomly chosen XY-coordinate. Nodes interact with each other by producing either a repulsion force, between non-connected nodes, or an attraction force, between connected nodes. By continuously updating the position of each node based on the sum of all of the forces acting upon it (net force), the nodes move around in the two-dimensional grid. The main result is that if a node has more edges within its own community than to other communities then the net force from all of the nodes in its community will pull it in it closer to its community and further away from other nodes.

Attraction Force: Two connected nodes produce an attraction force on each other. The strength of the attraction from v to u is: $\frac{1}{\text{degree(u)}}$. As the number of edges a node has approaches 1, the strength of its attraction force approaches 1.

Repulsion Force: Two non-connected nodes produce a repulsion force on each other. The strength of the repulsion from v to u is: $\left(1 - \frac{1}{\gamma(u,v)}\right)$. As the graph distance increases, the strength of the repulsion force approaches 1.

Distance and Force-Vector Strength: Regardless of whether a force from v to u is an attraction or repulsion force, it is linearly scaled down according to the euclidean distance between the positions of u and v in our two-dimensional grid. The closer two nodes are in the grid, the greater the strength of the force-vector between them. The strength of the vector from v to u is scaled by $\frac{M - \|\rho(u) - \rho(v)\|}{M}$, where M is the maximum euclidean distance in the two-dimensional grid.

This is an important calculation in the simulation. Without taking distance into consideration, only a few communities will ever be formed; as the force created by the largest community will overpower the forces from the smaller communities, compressing together two or more legitimate communities.

Our two-dimensional grid wraps in both the horizontal and vertical directions. Therefore we consider two forces between any two nodes, a shorter distance (dominant) and a longer distance (submissive).

Updating Node Positions: We update the position of node u as follows. Let Attr (a, b) be the force from node a to an adjacent node b and let Rep (c, d) be the force from a node c to a non-adjacent node d. Then the XY-position of node u, $(\rho(u))$ (where + and − are understood to be vector addition and subtraction), at time $t + 1$ is:

$$\rho^{t+1}(u) = \rho^{t}(u) + \alpha \times \sum_{v \in N,(u,v) \in E} \text{Attr}(v,u) - \sum_{v' \in N,(u,v') \notin E} \text{Rep}(v',u)$$

The α term acts as a multiplier for the attraction forces. In general we have found that α should be much larger than 1 to discover good community structures.

5.2 Convergence

Our algorithm does not always converge to an optimal community structure; for example, consider a node v that is closely related to two disjoint communities X and Y. In the optimal community configuration, node v is with community X but in this particular run it randomly starts closer to many nodes which belong to community Y.

As forces are dependent upon the euclidean distance between the nodes involved, it is possible that node v will get absorbed by community Y.

However, we find that majority of the solutions produced in our simulation have a high degree of similarity. Usually all of the community structures produced are variants of each other, capturing the strongest parts of the community structure.

5.3 Motivation

Recall that a configuration $\rho : N \rightarrow (\mathbb{R},\mathbb{R})$ assigns a two-dimensional euclidean position to all of the nodes in network G. Now consider the true community partition P^* for a given network $G = < N, E >$. The idyllic goal of our algorithm is to find the configuration ρ^* such that all of the nodes in the same community in P^* have the same position and all communities are as far apart as possible. Formally, ρ^* minimizes the following expression:

$$L(P^*) = \sum_{u \in N} \left(\sum_{v \in C_u} \| \rho(u) - \rho(v) \| - \sum_{v' \notin C_u} \| \rho(u) - \rho(v') \| \right),$$

where C_u is the community in P^* that contains u.

The obvious problem with determining $L(P^*)$ is that we do not know P^*. Instead we minimize an alternative expression, which offers an approximation to $L(P^*)$:

$$F(\alpha_G) = \sum_{u \in N} \left(\alpha_G \times \sum_{v \in E_u} \frac{\| \rho(u) - \rho(v) \|}{| E_u |} - \sum_{v' \in E_u} \| \rho(u) - \rho(v') \| \times (1 - \frac{1}{\gamma(u,v)}) \right),$$

where E_u is the set of nodes adjacent to node u, α_G is a scalar constant for this particular graph G, $| E_u |$ is the degree of node u and $\gamma(u, v)$ is the graph-distance as defined earlier. Considering our two earlier observations on properties of nodes within a community we claim that $\left(\frac{1}{|E_u|}\right)$ and $\left(1 - \frac{1}{\gamma(u,v)}\right)$ act as corrective terms which adjust the attraction and repulsion forces wrongly applied between nodes.

Consider two nodes u and v such that u and v are connected but are not members of the same community and thus (u, v) is an inter-community edge. In our model, there will be an attractive force between these two nodes, which could incorrectly label these two nodes as belonging to the same community. *Observation (i)* tell us that these inter-community edges are likely to be attached to a higher than average degree node (a potential hub). When calculating the strength of the force from node v to node u, we include the degree of node u by scaling the force by $\left(\frac{1}{|E_u|}\right)$. Thus, if u has a large degree, the force exerted on node u by its adjacent nodes will be reduced and including both nodes in the same community can be avoided.

Now consider the opposite situation where we have two nodes u' and v' that are not connected but are in the same community. These two nodes will produce a repulsion force on each other which could lead the two nodes to be far apart and thus mislabeled as belonging to two different communities. *Observation (ii)* tells us that if u' and v' are in the same community then the graph distance, $\gamma(u',v')$, between them is probably lower than the average graph distance. By scaling the strength of the force from node v' to node u' by $\left(1-\frac{1}{\gamma(u',v')}\right)$ we remove some of the prejudice lacking an edge between u' and v' has on our system.

5.4 *Clustering*

After updating the positions of each node, we run a clustering algorithm that updates the positions of the cluster centers. The clustering algorithm is based on Growing Cell Structures [5]. The main job of the clustering layer is to recognize the spatial separation between different communities.

1. Initially we create β cluster centers at random points in our two-dimensional grid.
2. We take each updated node position and assign it to the closest cluster.
3. We update the position of the clusters for time $t+1$, based on data at time t and on the position of the nodes that each cluster has won, as follows:

$$\rho^{t+1}(c)=\rho^{t}(c)+\frac{1}{2}\times\frac{1}{|\omega(c)|}\times\sum_{u\in\omega(c)}\|\rho^{t}(c)-\rho^{t}(u)\|,$$

where $\omega(c)$ is the set of nodes in cluster C.

4. Every λ time steps, we use the position of the cluster center h with the highest error as the initial position for a new cluster center.

$$\text{error(h)}=\sum_{\text{u}\in\omega(\text{h})}\|\rho(\text{h})-\rho(\text{u})\|$$

5. If the number of nodes that a cluster center wins drops below a set threshold, ε, for ξ iterations then that cluster center is removed and the nodes of that cluster will be absorbed by other clusters.

At the end of the simulation the cluster memberships represent the final community structure.

5.5 *Example*

Consider a simple network, G1, in Fig. 4. G1 has 8 nodes and 13 edges. There are two clear communities (which are visualized with circles and squares respectively) both are complete subgraphs of G1 with an associated $Q=0.423$.

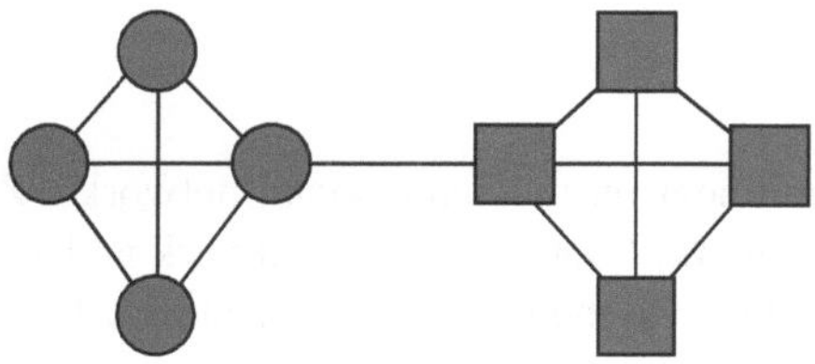

Fig. 4 A network, G1, with two complete subgraphs, each with four nodes. The two subgraphs are connected by a single edge

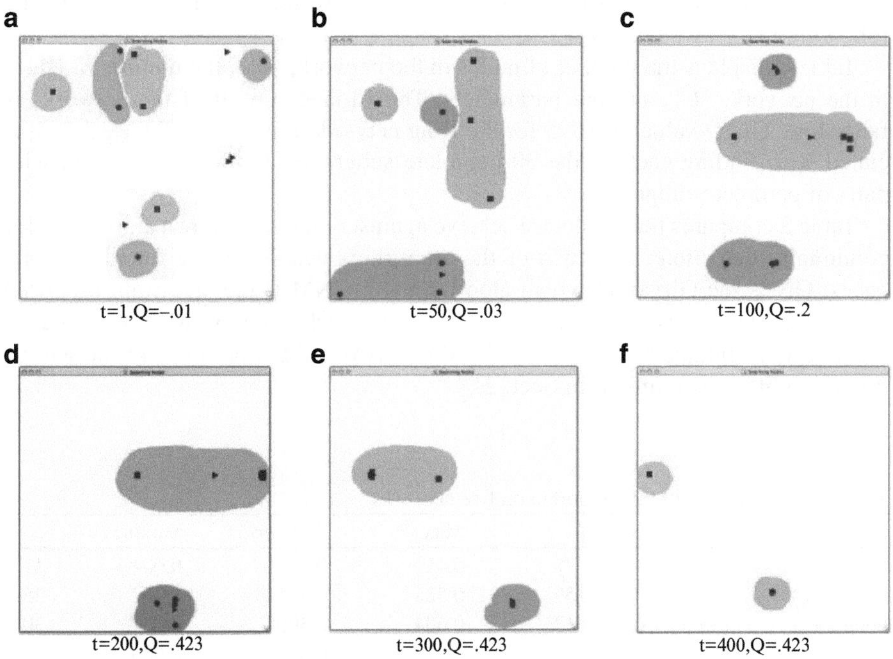

Fig. 5 The position of nodes and cluster centers for the network G1 as time increases. *Triangles* represent cluster centers, while *circles* and *squares* represent the two complete subgraphs. The modularity, Q, and current time, t, is given

In Fig. 5a–f, we display snap shots of our simulation from six different time-steps. Figure 5a shows the initial random positions of the eight nodes and three cluster centers (black triangles) at $t = 1$. The progression from Fig. 5a to Fig. 5f shows that the nodes in the two communities coalesce into two tight groups. The actual positions of the two clusters are not important, only the relative position of the two clusters. Shading around nodes is given only as a visual aid.

6 Results

We have run our experiment on the same social networks: Zachary Karate Club, Dolphin Interaction network, Football Game Network and Jazz Musician network as described earlier along with the ring described in Fig. 1.

6.1 Social Network Results

For each network we have run our experiment 50 times. The average Q-value, the highest Q-value, and the variance over Q-values are reported in Table 1. In the following results we take $\beta = 3$, $\xi = 4$, $\lambda = 5$, $\varepsilon = 4$.

Table 1 displays the number of nodes in the network, $|N|$, the number of edges in the network, $|E|$, and our parameters. The variance on all of the networks is fairly low. The Q-value of 0.875 for the Ring network, reported in Table 1, is associated with finding each of the 30 complete subgraphs, 0.888 is associated with pairs of complete subgraphs.

Table 2 compares the results we achieve against other well known algorithms for community detection[2]. For each of the algorithms details can be found in references: GN is the Girvan-Newman algorithm [8], CNM is the algorithm proposed by Clauset [1], EO uses the extremal optimization technique [3], MM is the matrix modularity technique by Newman [14]. MM, CNM and EO optimize Q and will be limited by its constraints from Sect. 2.

Table 1 Performance of our algorithm on five networks

Network	$\|N\|$	$\|E\|$	Max	Average	Variance	α
Karate	34	78	0.419	0.416	0.004	11
Dolphins	61	159	0.528	0.525	0.002	35
Football	115	613	0.604	0.590	0.004	30
Ring	150	630	0.875	0.875	0.000	15
Jazz	198	2,742	0.443	0.438	0.0001	60

Table 2 Comparison of our algorithm with four well known algorithms

Network	GN	CNM	EO	MM	This paper
Karate	0.401	0.381	0.419	0.419	0.419
Dolphin	0.38	NA	NA	NA	0.52
Football	0.60	0.57	0.60	NA	0.60
Jazz musicians	0.405	0.439	0.445	0.442	0.443

[2]Results for other algorithms shown in Table 2 are from [14] and [17].

Table 2 shows that our algorithm does well against other algorithms for these four networks. Furthermore, our algorithm does not use the Q metric, thus we are also able to find the correct community structure in the ring network, something no algorithm based on Q optimization can do.

7 Improvements

In the standard version of our algorithm both α and ε need to be user-supplied to the system to find a useful community structure. To remove these constraints, we introduce an information feedback loop from the clustering layer to the force-directed layer to automatically determine the values of α and ε.

7.1 α-Searching

We begin searching out useful values of α by noting the relationship between the amount of error in each cluster and the quality of the community structure. Our ideal solution would have each node located in its proper community with one cluster center at each community location (minimizing intra-cluster distance)[3]. Furthermore, we prefer each cluster center as far away as possible from the other community cluster centers (maximizing inter-cluster distance). We want to find an α value that gets our system as close to this ideal situation as possible.

We begin our simulation by setting $\alpha = N$, at each iteration we lower the α value by 1. When $\alpha \approx N$, the nodes will move around the screen in a chaotic manner, as α will be too high for communities to be found in the general case. However, as α decreases, nodes begin to settle down and are attracted to their most adjacent neighbors. At each iteration we also keep track of the global error averaged over all of the cluster centers in the system. When α reaches 1, we allow the α value to be adjusted by δ_α. While the error rate is decreasing, α is continuously moved in one direction. Once the error rate increases for Θ iterations, the direction that δ_α is applied reverses. Neither Θ nor δ_α are based on upon the size of the network. During this stage of α searching we restrict α to the range $\left[0, \frac{N}{2}\right]$.

7.2 Refining ε

The ε value is the minimum size for a set of nodes to be considered as a community. Instead of assigning a predetermined minimum number of nodes, it would be much better if we could dynamically determine the size of each community

[3]In general this is only possible with well-structured modular networks.

(in general we assume that this value should be greater than 2 for all networks). Here we point out that the good cluster centers repeatedly win a stable set of reoccurring nodes. For this reason we require nodes to maintain 75% of the same set of nodes:

$$\text{Stability(c)} = \frac{|\omega^t(c) \cap \omega^{t+1}(c)|}{|\omega^t(c) \cup \omega^{t+1}(c)|}$$

Any cluster c which does not win more than two nodes or has a stability <0.75, is removed from the simulation.

8 Results

We have re-run our improved algorithm on the same set of networks to compare the results with no user supplied data to the previous version where α and ε were required ahead of time ($\delta_\alpha = 1$ and $\theta = 4$ for all experiments).

Table 3 shows that the results using this feed back loop are comparative to the results obtained when using a tuned α and ε variable. However, it should be noted that the variances have increased and the averages have slightly dropped.

9 Conclusion & Future Work

The runtime of our simulation is in the order of $(S \times |N|^2)$, where S is the number of iterations to run the simulation (not including the time it takes to compute the graph distance between all pairs of nodes). This should be reduced in order to test our algorithm on large networks of tens of thousands of nodes. One way we suggest to address this issue is to create super-nodes during simulation, which will represent multiple nodes, thus reducing the total number of forces that have to be calculated. Super nodes will only be formed when we have a high degree of confidence that all nodes belong in the same community. Many successful runs of our algorithm stabilized to optimal or near-optimal configurations using the α, ε-improvements in only a few hundred iterations. An improved stopping criterion will allow the simulation to end earlier thus improving the run time complexity.

Table 3 Improved version on five well known networks

Network	Max	Average	Variance
Karate	0.419	0.400	0.059
Dolphins	0.526	0.500	0.24
Football	0.603	0.590	0.010
Ring	0.875	0.876	0.0001
Jazz	0.443	0.373	0.032

To summarize, we have presented an algorithm which makes use of no utility function in its calculations to determine the community structure. This simple algorithm provides competitive results against leading alternative algorithms. We have also shown that our algorithm can provide better results against specific limitations of Q.

References

1. Clauset A, Newman MEJ, Moore C (2004) Finding community structure in very large networks. Phys Rev E 70:066111
2. Davidson R, Harel D (1996) Drawing graphs nicely using simulated annealing. ACM Trans Graph 15(4):301–331
3. Duch J, Arenas A (2005) Community detection in complex networks using extremal optimization. Phys Rev E 72:027104
4. Eades P (1984) A heuristic for graph drawing. Congr Numer 42:149–160
5. Fritzke B (1994) Growing cell structures – a self-organizing network for unsupervised and supervised learning. Neural Netw 7(9):1441–1460
6. Fortunato S, Barthélemy M (2007) Resolution limit in community detection. Proc Natl Acad Sci USA 104:36–41
7. Fruchterman TMJ, Reingold EM (1991) Graph drawing by force-directed placement. Software Pract Exp 21(11):1129–1164
8. Girvan M, Newman MEJ (2002) Community structure in social and biological networks. Proc Natl Acad Sci USA 99:7821–7826
9. Gleiser P, Danon L (2003) Community structure in jazz. Adv Complex Syst 6:565
10. Kamada T, Kawai S (1989) An algorithm for drawing general undirected graphs. Inform Process Lett 31(1):7–15
11. Lusseau D (2003) The emergent properties of a dolphin social network. Proc R Soc Lond B 270(2):186–188
12. Newman MEJ, Girvan M (2004) Finding and evaluating community structure in networks. Phys Rev E 69(2):026113
13. Newman MEJ (2004) Fast algorithm for detecting community structure in networks. Phys Rev E 69:066133
14. Newman MEJ (2006) Modularity and community structure in networks. Proc Natl Acad Sci USA 103:8577–8582
15. Noack A (2004) An energy model for visual graph clustering. In: Proc. symposium on graph drawing (GD 2003). LNCS, vol 2912, pp 425–436. Springer
16. Noack A (2007) Energy models for graph clustering. J Graph Algorithm Appl 11(2): 453–480
17. Pons P, Latapy M (2006) Computing communities in large networks using random walks. J Graph Algorithm Appl 10(2):191–218
18. Radicchi F, Castellano C, Cecconi F, Loreto V, Parisi D (2004) Defining and identifying communities in networks. Proc Natl Acad Sci USA 101(9):2658–2663
19. Reichardt J, Bornholdt S (2004) Discovering functional communities in dynamical networks. Phys Rev Lett E 93:218710
20. Tasgin M, Bingol H (2006) Community detection in complex networks using genetic algorithms. ECCS Oxford, UK
21. Zachary WW (1977) An Information flow model for conflict and fission in small groups. J Anthropol Res 33:452–473

Comparing Two Sexual Mixing Schemes for Modelling the Spread of HIV/AIDS

Shah Jamal Alam and Ruth Meyer

Abstract In this paper we compare the impact of two sexual mixing schemes on the characteristics of the resulting sexual networks and the spread of HIV. This work is part of our studying social complexity in the Sekhukhune district of the Limpopo province in South Africa. While the agent-based models are constrained by evidence wherever possible, little or no evidence is available about individuals' choice of partners in the region and their sexual behaviour. Since we therefore have to depend on plausible assumptions we decided to study different sexual mixing schemes and their effect on the formation of sexual networks. We report on some fundamental network signatures and discuss the resulting HIV/AIDS prevalence as a macro-level output of the simulation.

Keywords Heterosexual networks • HIV/AIDS spread • Agent-based modelling

1 Introduction

The structure of sexual networks is critical for understanding the diffusion of sexually transmitted diseases (STDs) like HIV/AIDS. Sexual interaction among heterosexual individuals is regarded as the primary cause of high HIV prevalence in Sub-Saharan Africa [7,16]. However, very few empirical studies exist that report on complete sexual networks in an African community and study its features. The study by Helleringer and Kohler [7] on Likoma Island, Malawi, is one such exception. The topology and dynamics of a sexual network depend upon the cultural context of the respective society. Tendency of clustering, mixing behaviour and the variation in the sexual contacts are important determinants of the network and have significant implications in the spread of the STDs (c.f. [12]).

S.J.A lam (✉) and R. Meyer
Centre for Policy Modelling, Manchester Metropolitan University Business School, Manchester,M 13G H,U nitedK ingdom
e-mail:s hah@cfpm.org;r uth@cfpm.org

K. Takadama et al. (eds.), *Simulating Interacting Agents and Social Phenomena: The Second World Congress*, Agent-Based Social Systems 7,
DOI 10.1007/978-4-431-99781-8_5, © Springer 2010

This paper discusses two feasible mechanisms to model the formation of a sexual network. Each of these sexual mixing schemes comprises of a set of rules and assumptions that characterize sexual interactions among individuals in a simulated society. We have applied these schemes in the evidence-driven agent-based models developed in the context of a case study, which aims to understand the socioeconomic impact of HIV/AIDS in the village of Ga-Selala, in the Limpopo Province, South Africa [23]. The two models are a declarative model and a procedural model. The procedural model uses Repast[1] while the declarative model represents agents with logic-like rule bases and therefore combines Repast with the JESS[2] inference engine. A detailed description of the procedurally implemented model can be found in previous work [1,2]; its revised epidemiological component is discussed by Alam et al. [3]. Moss [14] describes key aspects of the declarative model.

In terms of outputs, procedural models naturally yield numerical series whilst declarative models yield both numerical series (usually at aggregated levels) and qualitative data at individual level. The question we address in this paper is whether different implementations – in our case declarative vs. procedural models – give qualitatively different model results. In the case of the models reported in this paper, we present initial comparisons concerning changes in the sexual networks and the prevalence of HIV/AIDS over time.

In the following sections, we compare the two sexual mixing schemes implemented in the procedural and declarative agent-based models. As the population changes over the course of the simulation, so do the sexual network characteristics. HIV is spread both horizontally (partner to partner) and vertically (mother to child) in the two models.

2 Specification for Two Sexual Mixing Schemes

Modelling dynamic social networks requires a mechanism for sexual mixing of the agents. Sexual behaviours and preferences for choosing a sexual partner depend heavily upon the social norms of the community. For our case study, we have implemented two sexual mixing schemes: one based on agents' aspiration and quality,[3] the other based on the endorsement mechanism as discussed by Moss [13]. We assume a traditional society where males take the first step in a sexual partnership. Thus male agents look for potential partners whereas female agents may accept or reject the courtship offers. In both the schemes, incest and inbreeding are prohibited.

The formation of sexual networks takes into account that people may have several concurrent sexual partners. The only exception is married female agents,

[1]Repast Agent Simulation Toolkit – http://repast.sourceforge.net/.

[2]JESS (Java Expert System Shell) – http://www.jessrules.com/jess/.

[3]This scheme is adapted from the mechanism originally proposed by Todd and Bilari [20].

who do not take part in extra-marital sexual relationships. Since same-gender partnerships are more or less taboo in the case study area the models only consider heterosexual relationships. The resulting sexual networks are therefore two-mode networks with males and females forming the two distinct sets of nodes.

2.1 A Scheme Based on Simple Aspiration and Quality Measure

In this scheme, we assign aspiration and quality ("attractiveness") log-normally to agents when they are created (cf. [8,19]). If the number of current partners is lower than his upper limit a male agent of age between 16 and 55 looks for a female between 14 and 40 whose attractiveness exceeds his aspiration level. He then sends her a courtship offer. Female agents evaluate all offers received during the current time step. All offers from agents whose attractiveness is below their own aspiration level are rejected immediately. From the remaining suitors a female agent will choose the best and accept his offer. The criteria for female agents to determine the best proposers changes with age [1]. For instance, young female agents prefer males of similar age, while more mature female agents may prefer unmarried employed suitors.

Agents search for sexual partners mostly among their friends and acquaintances. The friendship network is dynamic and based on the rules outlined by Jin et al. [9] for evolving clustered networks. There is also 5–10% chance for picking a female as potential partner at random. As the agents' search is dependent upon their friends and friends-of-friends circles, there is high likelihood that the potential partner is of similar age.

Agents without partners have their aspiration level successively decreased after a particular waiting time. For those satisfied with their current sexual partner(s), the aspiration level is updated incrementally. This is a simplified version of the original aspiration and choice criteria described in Simão and Todd [19].

From anecdotal accounts of the villagers we know that it takes about 1–2 years before a couple gets married. Since a male has to offer *lobola* (bride price) to the female's household before marriage, the marriage of the couple may be further delayed [2]. We initialize each agent with a maximal courtship duration sampled from a log-normal distribution with the minimum and maximum cut-off values set to 12 and 36 months. The courtship duration time (i.e. the time for dating until an agent decides to marry) is updated with age.

2.2 A Scheme Based on Endorsements

The declarative model makes use of so-called endorsements and individual tags in the agents' decision making. Endorsements can be thought of as labels used by an agent to describe certain aspects of other agents.[4] The model incorporates both positive labels like is-kin, is-neighbour, same-church, is-friend, similar, reliable, capable and negative labels like unreliable, incapable or untrustworthy. Some endorsements are static in that, once applied, they don't change over the course of the simulation (e.g. is-kin), while others are dynamic and may be revoked or replaced according to an agent's experiences [22]. All agents use the same list of endorsements but differ in how they assess them.

To do so, agents rely on a so-called endorsement scheme which associates each label with a weight to express how much store an agent sets by this particular aspect of a person. Weights are modelled as integer numbers between 1 and n for positive labels and −1 and $-n$ for negative labels, respectively. This allows for computing an overall endorsement value for a person, applying the following formula:

$$E = \sum_{w_i > 0} b^{w_i} - \sum_{w_i < 0} b^{|w_i|}$$

where b is a number base and w_i the weight of the ith label. Agents are assigned random endorsement schemes at creation, which differ not only in the weights they assign to the labels but also in the values used for n and b.

The overall endorsement value allows an endorsing agent to choose the preferred one(s) among a number of endorsees. To form the friendship network, for example, agents first determine which other agents are similar to themselves. This is based on abstract tags (to model character traits), which are assigned randomly to agents at creation. These tags are used to compute a similarity index (the number of similar tags), which in turn is used to generate "similar" and "most-similar" endorsements. Agents then compute the endorsement value for all known other agents and choose the ones with the highest values as their friends.

In case of the sexual network, endorsement values are used to model the attractiveness of potential partners. A male individual finds a female attractive if (1) she is of the same age or younger and (2) her overall endorsement value is higher than a certain threshold particular to the male individual. We thus assume that partner choice is biased by a preference for particular attributes.

Adults between 15 and 64 are considered to be sexually active. Males of this age group look for potential partners among the siblings of friends, work colleagues and the social groups they belong to. When they encounter a female adult they find attractive (see above) they make a pass at her, modelled as sending her a partner request.

Females evaluate all requests received at the same time and pick the best of the "applicants" if (1) their overall endorsement value is higher than the female's threshold and (2) in case they already have a certain number of current partners, if this applicant's endorsement value is higher than the lowest endorsement value of the current partners. In this case, the current partner with the lowest endorsement value is dropped, thus ending their sexual relationship.

[4]Endorsements were first devised by Paul [6] as a device for resolving conflicts in rule-based expert systems. Scott [13] modified and extended their use within a model of learning by social agents. This latter version of endorsements has been adapted for the declarative model.

Relationships may otherwise end due to marriage of one of the partners or – in absence of detailed knowledge about real reasons – are broken up randomly. The probability applied is influenced by the number of current partners and the person's age; it is highest for young people with a maximum number of concurrent partners.

3 Simulation Results

In this section we compare the results from the two sexual-mixing schemes introduced above: the first based on aspiration and quality (scheme-1) and the second based on endorsements (scheme-2). A typical run for each scheme is presented comprising of 35 (simulation) years. We apply typical network signatures to compare the structure of the resulting sexual networks and look at HIV prevalence and population size as relevant macro-level measures.

Scheme-1 is part of the procedurally implemented model using Java/Repast while scheme-2 is part of the declarative model, which also makes use of Jess. Both models use the probabilities of HIV transmission that are adapted from Wawer et al. [21]. The chance for mother-to-child HIV transfer was assumed as 30% [15]. Finally, both models were initialized with detailed household composition data derived from surveys in the Sekhukhune District, South Africa [17].

3.1 Characteristics of the Simulated Heterosexual Networks

Heterosexual networks are by construction two-mode and exhibit distinctive signatures different from other social and sexual networks. Typical signatures for heterosexual networks include spanning-tree like structure, large geodesic distance and diameter for any two connected pairs of nodes, low density and low clustering [18]. Another important characteristic of heterosexual networks is the existence of very few cycles (i.e. 4-cycles and cycles of higher order).

As the study of an empirical heterosexual network [5] shows, spanning-tree like structures and few cycles reflect the influence of a bias in partner selection, which is not observable in random mixing schemes [10,11]. Large geodesic distance and diameter of the network also reflect that a majority of the individuals (nodes) may not be aware of others' sexual links in the population. In the case of HIV spread, being linked in a long chain of people can drive the prevalence much faster.

Figure 1 shows snapshots of the sexual networks resulting from the two schemes taken in the middle and towards the end of the respective simulation runs. As can be seen, the networks from both schemes exhibit the characteristic spanning-tree like structures and few cycles. In addition, a number of 3-stars and dyads can be found. 3-Stars are indicative of promiscuous behaviour (as are the spanning-tree like characteristics of the larger components) while dyads may include married couples where the male was unable to find another female sexual partner.

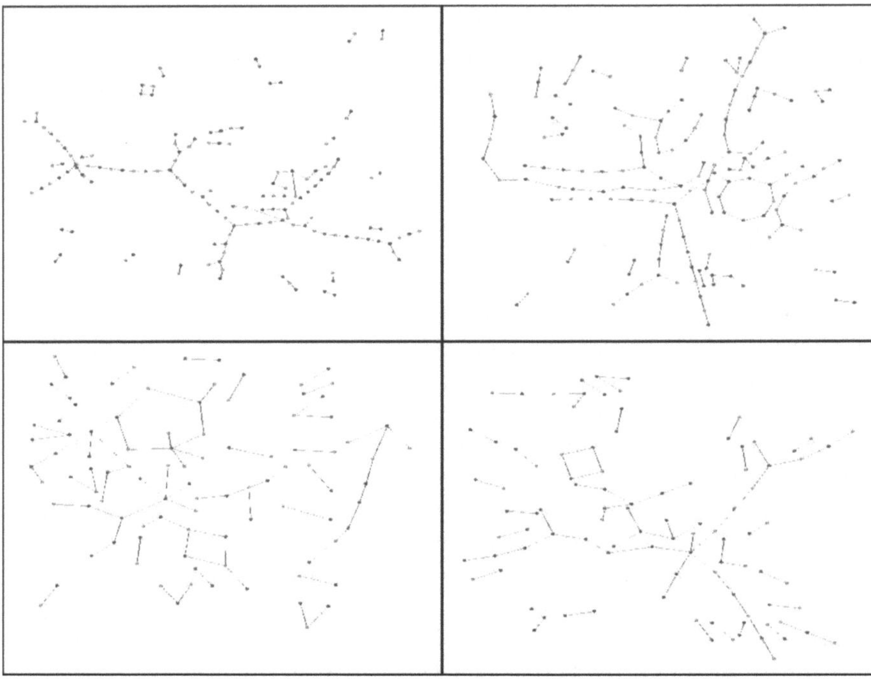

Fig. 1 Snapshots of the heterosexual networks resulting from scheme-1 (*top*), taken at the 21st and 33rd year, and scheme-2 (*bottom*), taken at 21st and 35th year

To compare the differences of the two simulated networks we calculated some basic characteristics for every sixth month using Pajek.[5] Membership in the simulated sexual networks requires the individuals to have at least one sexual partner (married or unmarried) at a particular time step.

Figure 2 (top) shows the overall density for both networks. The decline in network density for scheme-1 coincides with the decline in the overall population size (cf. Fig. 4; bottom). On the other hand, the stable population in the first 20 years for scheme-2 enables sexually-active agents to form more concurrent partnerships than in scheme-1. The weekly time step used in the declarative model (scheme-2) as opposed to the monthly time step in the other model (scheme-1) might influence this outcome as well. We will need to investigate this further.

Comparing the relative size of the largest network component depicts volatile effects in both cases (Fig. 2; bottom). However, scheme-1 shows a relatively higher proportion than scheme-2. This can be explained by a chance, albeit small, for male agents to randomly pick a female partner. This may result in merging two or more smaller components into one large connected component. A more significant measure of comparison would probably be the complete distribution of components' sizes, which we will aim for in future work.

[5]http://vlado.fmf.uni-lj.si/pub/networks/pajek/.

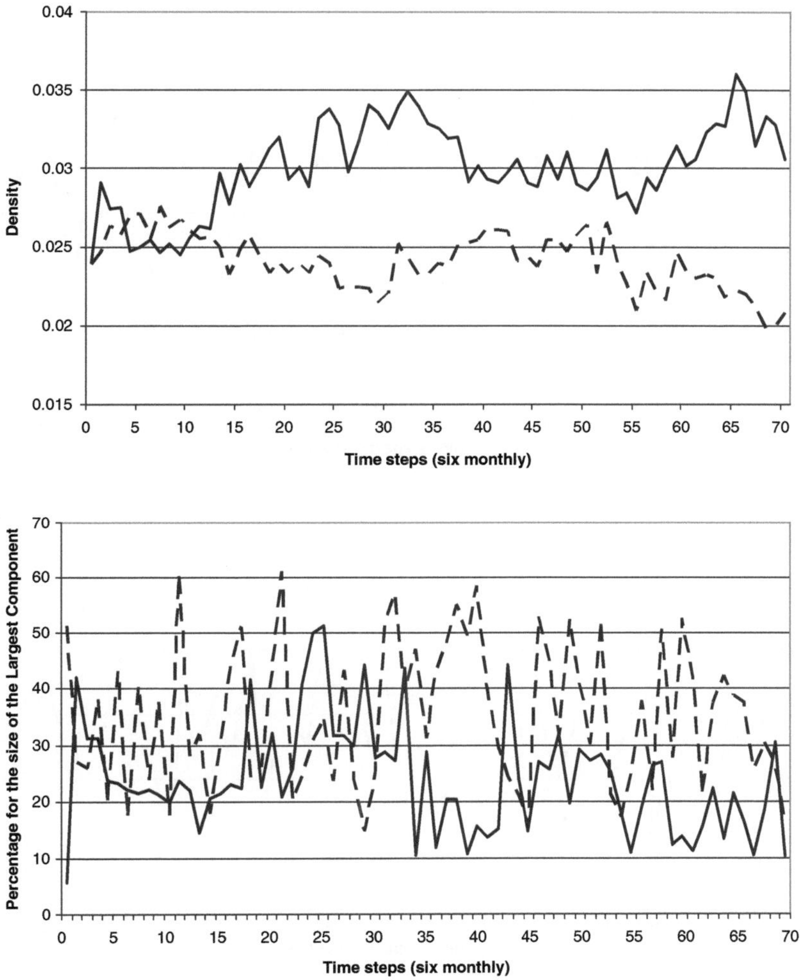

Fig. 2 Density (*top*) and percentage size of the largest component in the network (*bottom*); *dashed line* represents scheme-1; *solid line* represents scheme-2

Figure 3 (top) depicts the geodesic distance for any connected pair of individuals, whereas the bottom chart shows the diameter of the two networks. Unsurprisingly, scheme-2 exhibits a much lower geodesic distance due to the higher density of the network. Similarly, the diameter of the network for scheme-1 is generally higher since the size of its largest component is almost always greater than for the network of scheme-2. It shows that scheme-1 has more long 'chainlike' structures as observed by Bearman et al. [5], although we do not know if similar characteristics are found in our case study region as well. As both schemes do not assume a fixed population size, the size of the network is also a major influence on the outcome of these network statistics.

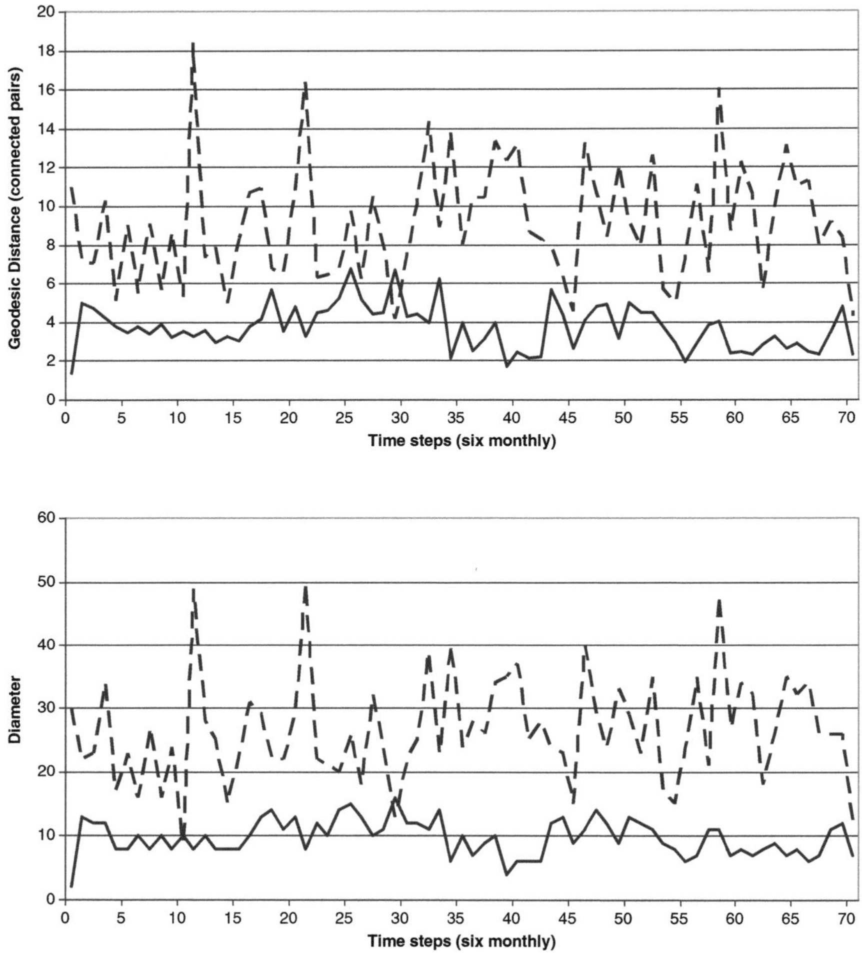

Fig. 3 Geodesic distance (*top*) and diameter of the networks (*bottom*); *dashed line* represents scheme-1 and *solid line* scheme-2

3.2 HIV/AIDS Prevalence

As can be expected, the different network structures arising from the two different sexual-mixing schemes have different outcomes on the macro level. It is interesting to observe clustered volatility in the time series of the network characteristics discussed in the previous section (cf. Figs. 2 and 3). This effect, however, is not significant in the prevalence of HIV/AIDS and the population size as shown in Fig. 4.

Figure 4 (top) shows the prevalence of HIV/AIDS over 35 years for the two sexual mixing schemes at a monthly scale. In both cases, a number of agents have been initialized as HIV-positive at the start of the simulation, based on the current

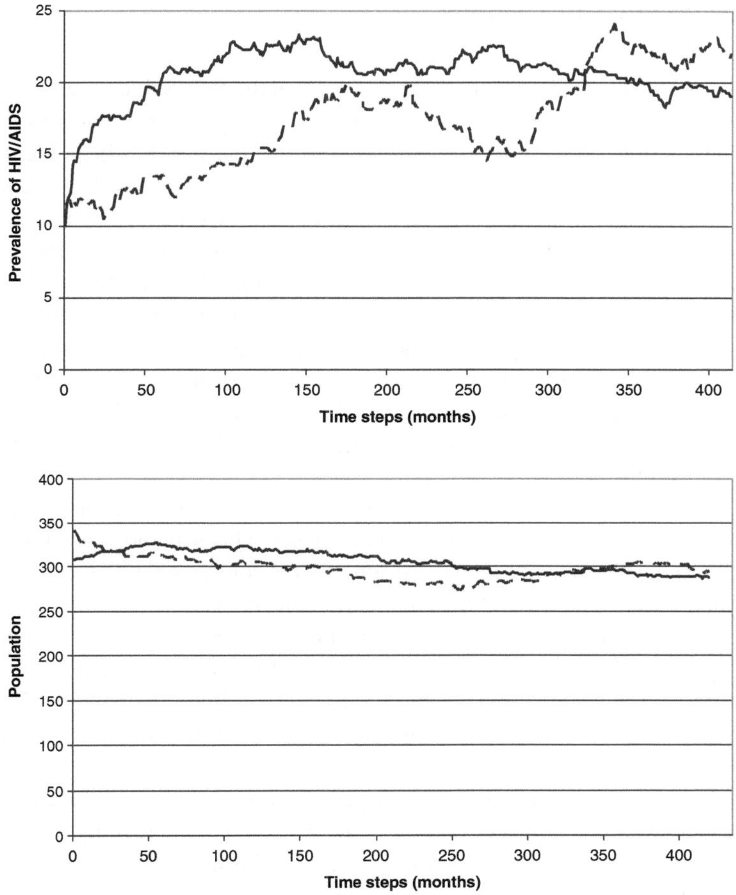

Fig. 4 HIV/AIDS prevalence (*top*) and population size (*bottom*) over 35 years for a typical simulation run; *dashed line* represents scheme-1, *solid line* scheme-2

HIV/AIDS prevalence in the case study region. While in the case of scheme-1 prevalence progresses slowly it quickly reaches about 23% in the case of scheme-2. This can be explained by the higher density of the sexual network and the time scale of the declarative model, where new cases of HIV (via sexual transmission per coital act and external incidence among migrants) are introduced at a weekly basis. After a short decline the prevalence reaches a plateau of about 20% around the 15th year (month 180). In contrast, with scheme-1 the prevalence shows a significant drop between months 225 and 275 (year 18–23) and then climbs again.

Typically, the spread of an epidemic is characterized by the rise in prevalence, followed by a decline and then attaining equilibrium for some time. Scheme-2 shows this characteristic of the curve while scheme-1 shows more volatility. As both simulations were run for only 35 years, it would be interesting to investigate whether the prevalence stabilizes in the next 50 years for the two cases.

4 Discussion and Outlook

Evidence-driven modelling of the spread of HIV in the Sub-Saharan region requires extensive fieldwork, clinical trials and involvement of the local stakeholders. Sexual interaction among promiscuous heterosexual partners has been identified as the primary cause of HIV/AIDS prevalence in the region. This involves individuals' sexual preferences, migration and the role of commercial sex workers. In the absence of detailed empirical data, any individual-based modelling of the spread of HIV has to depend on plausible assumptions regarding the sexual behaviour of the population. It is therefore important to investigate the effects of a chosen model mechanism on the resulting sexual network and overall simulation outcomes. In this paper, we have briefly explored the implications of using two different sexual mixing schemes in agent-based models constrained by evidence wherever possible. This is part of our effort to develop our understanding of the underlying phenomena in a bottom-up fashion.[6]

The first sexual mixing scheme is based on a one-dimensional interpretation of 'attractiveness' and preference for a sexual partner. This scheme was chosen for its simplicity in the absence of relevant evidence.[7] It is contrasted with the contextually rich second scheme, based on the endorsement mechanism. Both schemes incorporate the notion of an individual's self-perception and their perception of a potential partner (c.f. [4]). A wide range of models from social psychology and social anthropology exist on mate selection. Due to constraints in both available evidence and resources, however, we have not yet integrated other partner selection mechanisms in our study.

Although this paper can only present a preliminary comparison of the two sexual mixing schemes, the initial results already give some interesting insights with regard to relevant parameters for both schemes. For instance, the relatively high and more stable prevalence in scheme-2 indicates that the average lifespan of HIV-positive agents is greater than in the other model. This can be explained by the declarative model determining the chances for an individual to die at a given age according to the WHO life tables (c.f. [14]), which base mortality rates on all known causes of deaths including AIDS. On the other hand, HIV-positive agents in the procedural model leave the system following the late infection stage. The results of the declarative model may thus refer to a situation where an anti-retroviral treatment is available; resulting in a greater lifespan for HIV-positive individuals.

Another important issue is the size of the simulated population in both cases. Typically, models of infectious diseases assume large population sizes so that the stochastic effects are negligible. The agent population in the two models presented

---

[6]See Moss [14] for further discussion on this subject.

[7]Elsewhere, Alam [1] present a more elaborate and a contextualized enhancement of the first sexual mixing scheme. The results presented in this paper are based on the scheme described in our previous work [3].

in this paper, however, was kept at the low scale of the population of a small village, which relates to the fluctuations in the characteristics of the simulated sexual networks at the individual level.

For our future work, we intend to initialize both models with the same sexual network. This will allow closer alignment of the two models and help to perform a detailed analysis of the two sexual mixing schemes.

Acknowledgments This work was done within the CAVES project, which was funded under the EU 6FP NEST programme. We are thankful to the editors, the reviewers and the participants of WCSS'08 for their feedback and comments. Thanks to Scott Moss, Bruce Edmonds, our colleagues at the Stockholm Environment Institute (Oxford) and at the Centre for Policy Modelling for their input and feedback over the course of the CAVES project and this work.

References

1. Alam SJ (2008) On understanding the social complexity in the context of HIV/AIDS: a case study in South Africa. Doctoral thesis, Center for Policy Modelling, Manchester Metropolitan University, Manchester, UK
2. Alam SJ, Meyer R, Ziervogel G, Moss S (2007) The impact of HIV/AIDS in the context of socioeconomic stressors: an evidence-driven approach. J Artif Soc Soc Simulat 10(4):7. http://jasss.soc.surrey.ac.uk/10/4/7.html
3. Alam SJ, Meyer R, Norling E (2009) A model for HIV spread in a South African village. In: Multi-agent-based simulation IX (MABS'08). LNAI, vol 5269. Springer, pp 33–45
4. Buston PM, Emlen ST (2003) Cognitive processes underlying human mate choice: the relationship between self-perception and mate preference in Western society. Proc Natl Acad Sci USA 100(15):8805–8810
5. Bearman PS, Moody J, Stovel K (2004) Chains of affection: the structure of adolescent romantic and sexual networks. Am J Sociol 110(1):44–91
6. Cohen P (1985) Heuristic reasoning – an artificial intelligence approach. Pitman, Boston
7. Helleringer S, Kohler H-P (2007) Sexual network structure and the spread of HIV in Africa: evidence from Likoma Island, Malawi. AIDS 21(17):2323–2332
8. Heuveline P, Sallach D, Howe T (2003) The structure of an epidemic: modelling AIDS transmission in Southern Africa. In: Proc. symposium on agent-based computational modelling, Vienna
9. Jin EM, Girvan M, Newman MEJ (2001) Structure of growing social networks. Phys Rev E 64:4
10. Koopman JS et al (1997) The role of early HIV infection in the spread of HIV though populations. J Acquir Immune Defic Syndr Hum Retrovirol 14:249–258
11. Kretzschmar M, Morris M (1996) Measures of concurrency in networks and the spread of infectious disease. Math Biosci 133(2):165–195
12. Liljeros F, Edling CR, Amaral LAN (2003) Sexual networks: implications for the transmission of sexually transmitted infections. Microbes Infect 5:189–196
13. Moss S (1995) Control metaphors in the modelling of decision-making behaviour. Comput Econ 8:283–301
14. Moss S (2008) Simplicity, generality and truth in social modeling. In: Proc. second world congress on social simulation (WCSS'08), Fairfax, VA. http://cfpm.org/cpmrep187.html
15. Newell M et al (2004) Mortality of infected and uninfected infants born to HIV-infected mothers in Africa: a pooled analysis. Lancet 364:1236–1243
16. Nordvik MK, Liljeros F (2006) Number of sexual encounters involving intercourse and the transmission of sexually transmitted infections. Sex Transm Dis 33(6):342–349

17. RADAR (2005) The RADAR IMAGE study – intervention with microfinance for AIDS and gender equity. http://web.wits.ac.za/Academic/Health/PublicHealth/Radar/
18. Robins G, Pattison P, Woolcock J (2005) Small and other worlds: global network structures from local processes. Am J Sociol 110(4):894–936
19. Simão J, Todd PM (2003) Emergent patterns of mate choice in human populations. Artif Life 9(4):403–417
20. Todd PM, Billari FC (2003) Population-wide marriage patterns produced by individual mate-search heuristics. Agent-based computational demography: using simulation to improve our understanding of demographic behavior. Physica-Verlag, New York, pp 117–137
21. Wawer MJ et al (2005) Rates of HIV-1 transmission per coital act, by stage of HIV-1 infection, in Rakai, Uganda. J Infect Dis 191:1403–1409
22. Werth B, Geller A, Meyer R (2007) He endorses me – he endorses me not – he endorses me ... contextualized reasoning in complex systems. In: Papers from AAAI Fall Symposium: emergent agents and socialities – social and organizational aspects of intelligence. AAAI technical report: FS-07-04
23. Ziervogel G et al (2006) Adapting to climate, water and health stresses: insights from Sekhukhune, South Africa. Stockholm Environmental Institute Report, Oxford, UK

Exploring Context Permeability in Multiple Social Networks

Luis Antunes, João Balsa, Paulo Urbano, and Helder Coelho

Abstract In agent-based social simulation, agents engage into several concomitant relations, forming social networks. In these several networks, agents are immersed in contexts, in which they contact more often with some agents than with others. Moreover, for the complex cases involved in social simulation, most of these networks will be multi-modal, comprehending networks of agents which form themselves other networks, and so forth successively, including self-organised and institutional aggregates. We have proposed the concept of permeability between contexts, and have showed how in certain conditions the explicit representation of these multiple relations can enhance the dissemination of phenomena in an important feature for the analysis of the complex dynamical behaviour of the networks. In this paper, we further explore this research by considering the concepts involved in the representation of these complex social descriptions, and other related concepts, such as roles. We compare the results of previous simulations with a variation of the consensus game, and conclude that the permeability between contexts in different relations can greatly enhance the dissemination of phenomena throughout the society, allowing to explore rich informational and dynamical content of the networks.

Keywords Context permeability • Social networks • Consensus game

1 Introduction

In multi-agent-based simulations designed to study complex social phenomena, agents are usually organised into some underlying structure that supports their contact, communication and interaction. To the purpose of facilitating simulation visualisation (which in turn is important for the experimenter in that it allows for a more intuitive view on the evolving patterns) most of the problems get represented

L.A ntunes(✉), J. Balsa, P. Urbano, and H. Coelho
GUESS/LabMAg/Universidaded eL isboa, Lisboa,Por tugal
e-mail: xarax@di.fc.ul.pt; jbalsa@di.fc.ul.pt; pub@di.fc.ul.pt; hcoelho@di.fc.ul.pt

K. Takadama et al. (eds.), *Simulating Interacting Agents and Social Phenomena: The Second World Congress*, Agent-Based Social Systems 7,
DOI 10.1007/978-4-431-99781-8_6,

into compressed systemic frameworks. This compression includes the reduction of the dimensionality involved, the use of abstraction to depict connections between agents, the uniformisation of the agent granularity, and other techniques. If this compression effort is adopted too soon in the modelling effort, it brings the risk of masking away the due complexity of the problems, and force the experimenter into solutions that are closer to his/her expectations (of what s/he can represent, of what can be sought after, of what can be allowed to come out of the simulations) than what the problem would demand for.

Whilst admitting that visualisation is an important desirable feature for social simulations of complex phenomena, we argue in favour of keeping the data as close to their original format for as long as possible. In this vein, we advocate the maintenance of richer representations of the complex set of relations in which the agents are involved in. These concomitant relations are closer to the representations traditional social scientists use in their research, and can be explored through simulation in a manner that can ultimately reveal so far hidden properties and phenomena. In particular, we are interested in exploring the permeability between contexts in these relations: how a particular phenomenon can disseminate in a qualitatively and quantitatively different manner through the communication that the permanence of multiple concomitant relations supports. Having in mind an abstract exploration of the potential of these ideas, we keep our definitions as clean and simple as possible: contexts are neighbourhoods in a given relation, and permeability (contact points between relations) is ensured by having the same agents present in several concomitant relations. For instance, consider two relations: family and work colleagues. Taking agent α, a context in relation family could be its spouse and children, whereas in relation work colleagues it could be composed by the agents sharing an office with α. Agent α ensures permeability between these two contexts.

In this paper, we further explore these ideas through the use of the new feature of NetLogo 4: links. Whereas relations would have an intricate representation in previous versions of this package, this new concept allows for a swifter and more transparent representation of relations between agents. We have tested out this framework through a very abstract game, the consensus game, so that not to overcommit to some application domain. Our results have showed that in certain circumstances, there is a qualitative difference in the structure of the convergence to a societal consensus. In this paper we show the results of experiments over the same games, but with the agents spanning over a variety of typical topologies.

The paper is organised as follows. In the next section we briefly address agent relations and their variety, focusing on the kind of social networks that are more frequent in the literature. Section 3 describes the idea of relations and contexts, and relates them to the concept of role. In Sect. 4 we argue in favour of simulation compared to mathematical or statistical analysis when addressing social networks data. The following section describes the consensus games we use to illustrate these ideas and examine their impact. Section 6 addresses the representation formalism used in the experiments, NetLogo 4 links. We finish the paper by mentioning the most important results of the experiments and drawing some conclusions.

2 Multiple and Multi-Modal Relations

In every relevant problem to be studied, agents will be involved in several concomitant relations. This complex setting gets only worse if we consider that even to define the relevant actors for a particular application may require a structure of intertwined levels of abstraction. When individuals are embedded in networks that are embedded in networks we have a "multi-modal" structure, such as students who together with a teacher form a classroom, of which in turn a school is composed of [8]. Surely the panorama can get more complicated, as these several networks of relations do not necessarily have such a "regular" meta-structure.

In an effort pointed at the study of decision change in agricultural networks, Amblard and Ferrand [2] propose a multi-agent system whose agents represent the model actors, the relations between them and the cliques formed by those actors. Actors are characterised by general, relational and decisional attributes. The relational attributes determine a "relational lacking" that can be considered a driving motivational force for the actor. If some general attributes seem to overdetermine too soon the structure of the simulation (e.g. the division of socioeconomical status in aristocracy, bourgeoisie and working class seems too restrictive even for an early XX century representation of a urban zone), this model possesses some self-reflective characteristics that render it quite interesting and general. The behavioural dynamics is based on relations, which allows for self-motivated agent re-structuring. This offers an alternative to data driven models [10], although possibly empirically less reliable. However, even data driven models must suffer from some model preconceptions since data necessarily have a source. The societal models organised into a complex ontology from isolated actor to domain-based systems, and including some time-aware models of relations is a good starting point for design explorations, since it is sufficiently detached from specific applications.

For the time being, in an attempt to control complexity and focus on the study of dynamic consequences of the topological structures underlying social simulations, we will opt for what we can call a "first order" approach. We will take actors and relations between them as givens of the problem, in a similar way as what is done in [12]. Here the authors select relations among scholars in a series of scientific conferences, namely *meets*, *knows*, and *collaborates*.

Our agents will be the atomic individuals of the simulations, but our relations will be kept abstract, so that we can concentrate on the dynamics they induce. We can think of the relations in our simulation as reasonably stable, such as family ties, or work colleagues, while we study the consequences of their mutual connectedness. In the final section we enumerate some of the possible ways to proceed this study into more complex scenarios, following an incremental deepening of the concepts recommended by the e*plore methodology [4, 9].

3 Relations, Roles and Contexts

The concept of role has been traditionally adopted in artificial intelligence to account for multiple engagement into several activities. An extensive survey of this notion can be found in [11]. Masolo et al. describe the account of roles adopted in several disciplines, including knowledge representation (role as a binary relation), knowledge engineering (roles as task ontologies representing individuals), object-oriented and conceptual modelling (roles as places in a relationship). The approaches that is more interesting to us is when roles are considered from a multi-agent system or a sociological/philosophical standpoint. In multi-agents systems, roles are seen as restrictions to behaviour, an abstract description of an entity's expected functioning. This assumes many times a deontic characterisation that includes time dynamics and a structure of dependencies and relations between roles and individuals that fulfil them. In the sociological approach, roles are seen as behaviours specific to a set of persons in a context, including sets of rights and duties, acting parts of expected patterns. Masolo et al. proceed to provide a first-order approach to the notion of social roles, which is quite interesting: roles are properties, can be predicated, roles are anti-rigid and have dynamic properties (temporally evolving, or more generally considering other modalities), roles have a relational nature, roles are related to contexts.

The notion of context has also remained a hot issue in the literature over the years. McCarthy proposed to clarify the notion some years ago [13], but did not go much further than an ambiguous idea involving some structure in which first-order formulas could be evaluated and related to each other. Contexts can be transcended, and evaluation can be relatively decontextualised. Contexts provide a referential basis for linguistic processing, and can be related to each other to provide concept synchronisation (for instance in databases).

Many of these concepts seem to be basilar for the others, and their definition and properties are far from consensual in the literature. The notion of social relation as naïvely described in the social networks literature seems too dry and simplistic. Roles do seem to suffer from an overly heavy deontic character that might render them attractive for a logician but impractical for the simulator. The logic underlying a usable theory of roles would have to include a complex structure of contexts to provide grounding of the concepts, behaviours, and intricate relations intra- and inter-agents (let alone them and institutions, since we are postponing multi-modality). That structure of contexts would constitute an ontological challenge in itself, with complete references to the symbols in the agent's mind, making it unbearably difficult to obtain even the simplest coherent behaviour. This is the type of difficulties that usually drives modellers into the use of simplistic theories such as utility theory, where numbers and friendly mathematics can make uniform and simple what is inherently and utterly complex.

In our case, simplicity (arguably excessive, or, as modellers defensively prescribe, necessary) comes from the eyes and experience of the modeller. It has been proven useful to use different researchers to accomplish the several phases of the development of simulations [4]. Different views from different disciplines can contribute to avoid the formal-conceptual prejudice that often computer programming approaches carry-[10]. Relations drawn by social scientists can help to gradually focus on a simpler world of representations, by proceeding towards the direction of answering the relevant research questions. Granted, perhaps the precocious shaping of those questions in the representational framework prevents other perspectives to be adopted, but hopefully the multi-disciplinary dialogue can open the problem before narrowing towards shaped solutions.

Generic arbitrary roles seem to carry additional disadvantages, such as some impermeability between each other, which implies lack of grounding with essential (corporeal/bodily/motivational) features of the agent that is currently fulfilling. This impermeability carries through several modalities, including time, which is fundamental to simulation. On the other hand, relations such as the ones found in social networks, can be subject to dynamical analysis (for instance, stochastic), can be built over themselves by drawing arbitrarily complex structures of relations deemed relevant to the simulation (as shown in [2]). Graph theory and equilibrium laws can be searched for through conventional techniques, but we are more interested in analysis aggregate behaviours together with individual trajectories, and the reasons for both [5].

4 Social Network Representations and Analysis

Most graph representations of real world social networks follow patterns that have only recently being revealed. Such is the case of scale-free networks [6]. Scale-free networks can be defined as a connected graph in which the number of links (k) connected to a given node follows a power law distribution, $P(k) \sim k^{-\gamma}$. Scale-free networks are to be found in a great variety of real situations, and display the property that most of the nodes have little connections, whereas some network nodes (usually called "hubs") are highly connected. This is depicted by a right-skewed, of "fat tail" distribution. Barabási and colleagues proposed a simple generative mechanism called "preferential attachment" (cf. [1]). Although these mechanisms only generate a subset of the universe of scale-free networks, this is what we used for our experiments, and with γ fixed to 3 (most real data exhibit γ on the range [2, 3], although sometimes, smaller exponents can arise [14].

Scale-free networks have the additional property of being "small-world networks" (although there are other small-world networks other than scale-free). This means that while most nodes are not neighbours to one another, they can still be connected by a small number of connections. Scale-free networks have other interesting properties, such as a close to constant diameter as the number of nodes grow ($d \sim \ln \ln n$), or a certain "fault-tolerant" behaviour (problems affecting random nodes will hardly fall on the critical hub nodes). In small-world networks, even though there is a high incidence of cliques (or subgraphs close to cliques), there prevails many times a popular notion (the famous "six degrees of separation" property) that it is easy to link any two people together by a path with only a few connections. While true in theoretical terms, this notion is based on the idea of connecting two nodes that have *some* kind of relation between them, when in reality the social world is much more complicated than that: people have all kind of relations linking them (family, work, acquaintances, etc.) and moreover *know* a lot about (what they know about) those relations. This is what often causes the "It's a small world!" utterance in real life, and must not be masked away by hasty simplifications on the modeller's side [3].

Most of the analysis of social networks is done in quite statical terms. Mathematics and statistics tools only start to provide the possibility of dynamical analysis [7]. Given the purpose of multi-agent-based social simulation, it is fundamental that useful dynamical properties, even some ones linking individual behaviours to global behaviours, can be derived from the network analysis. With our approach, we aim to contribute to such research endeavour, by bringing the fields closer together and feeding on each other. Our approach is to use simulation to explore the design space, not only of agents, but also of societies and even experiments. The key point in these simulations is to understand to what extent the structure of connections the agents engage in simultaneously can have a role in the shape of convergence towards a simple collective common goal, an arbitrary consensus.

5 Consensus Games

To try out some of the theories without commit ourselves to applications that could demand for shaping of the relevant relations we picked up a really simple example, a consensus game [3, 15]. In this case we select a variant of what is called the majority game. Each agent has a current choice towards one of two possible options (say, green and red), which are for all purposes arbitrary (no rationality or strategic behaviour here). Every time an agent *meets* another one, each of the agents has the chance to either keep its current choice, or change towards the other agent's choice. In the variant we use in these experiments, agents keep track of previous encounters and when engaged in a new encounter it calculates the total number of agents of each colour it has seen before and adopt the colour of the majority of those.

6 Experimental Setting

Results are taken over 80 simulations. In each simulation we have 100 agents. In each simulation, agents are deployed in a number p of "planes," representing different relations (or different views over a more complex relation, which is equivalent). Each agent is present in every plane, but its connections are possibly different in every plane.

Before the simulation starts, the agents are set up for the planes involved. Experiments are run with agents organised either over a "regular graph" or over a scale free graph. This initial set up determines whom the agents can have contact with. For the multiple-plane experiments, all agents are present in every plane (every relation), but are initially set up independently. We run simulations with one, two, and three planes, with all possible combinations of the following base agent distributions: k-regular (with $k = 1, 2, 3, 4, 5, 10, 20, 30, 40, 50$), and scale free (with $\gamma = 3$). For regular networks we run experiments with four planes. The parameter k-in regular networks means that each agent is connected with exactly $2k$ other agents. This means that for $k = 50$ the graph is fully connected.

In each cycle, we select 100 pairs agent-relation (i, j), with $i \in \{1, ..., n\}$ and $j \in \{1, ..., p\}$. So the same agent can get selected more than once in on cycle, if a different relation gets picked. Each agent in one selected pair (i, j) has one meeting (or encounter) with the agents that are directly connected to it. In this meeting, the agent plays the majority game: it updates its record of the total meetings it had with agents of each colour, and then its own colour is updated to the colour that has the majority. The update policy of the agents is sequential in the chosen 100 pairs. The simulation stops after 3,000 cycles (and so 300,000 meetings) or whenever total consensus is achieved.

7 Analysis of Simulation Outcomes

The following tables show the results of our simulations. Table 1 shows the percentage number of times that convergence was achieved in the 3,000 cycles. We can see clearly that for some networks (regular with small k and scale-free) consensus is never achieved in one single plane. However, as soon as we provide context permeability by adding more planes, consensus is achieved in a significantly greater number of occasions. These results are especially exciting for the scale-free networks, so common in real world domains: as soon as we add more planes, we can achieve consensus in a lot of occasions. For instance, with three planes we already have consensus in 82% of the cases. Figure 1 illustrates the results.

Moreover, Table 2 shows that to add more planes has a dramatic reduction effect in the total number of meetings necessary to achieve the consensus. This is true for all the networks except for those that are almost fully connected ($k \geq 30$) and for the fourth plane in some other cases. Scale-free networks display numbers similar to regular networks with small k (2 or 3), although convergence is quite slower for

Table 1 Percentage of times consensus was achieved with all planes equal in kind

$k =$	1 (%)	2 (%)	3 (%)	4 (%)	5 (%)	10 (%)	20 (%)	30 (%)	40 (%)	50 (%)	s-f (%)
1 pl	0	0	0	0	0	19	73	95	100	100	0
2 pl	5	65	90	90	93	98	100	100	100	100	34
3 pl	65	90	100	100	100	100	100	100	100	100	82
4 pl	100	95	100	100	100	100	100	100	100	100	94

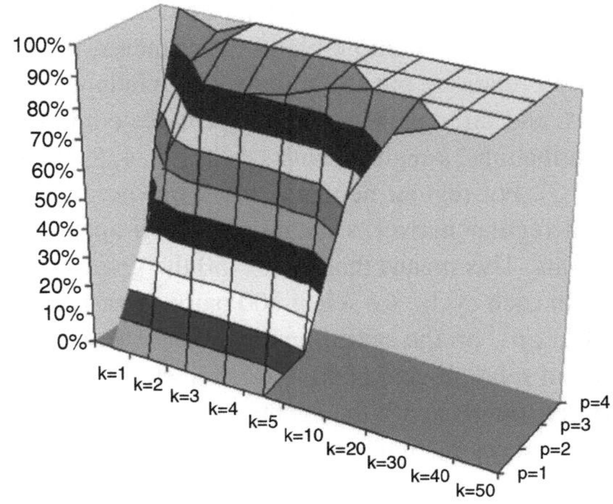

Fig. 1 Percentage of times consensus was achieved

Table 2 Average number of meetings to achieve consensus with all planes equal in kind

k =	1	2	3	4	5	10	20	30	40	50	s-f
1 pl						3,073	2,314	2,584	1,700	1,475	
2 pl	96,050	58,485	43,481	14,286	9,997	4,718	2,505	2,125	1,895	1,950	58,196
3 pl	50,746	6,839	3,845	3,640	3,210	2,530	2,317	1,861	1,950	1,911	22,035
4 pl	21,070	9,711	5,025	2,895	3,230	1,960	1,812	2,166	2,054	2,044	13,353

more than two planes. We should note however that convergence is achieved in a rather reasonable amount of meetings, even in the worst cases, while in the vast majority of cases the number of meetings in very low. Figure 2 gives us a visual grasp of these results.

On top of these experiments with several planes, we also run experiments with combinations of scale-free with regular networks. So, for two planes we run simulations with a regular network and a scale-free network, while making k span over 1, 2, 3, 4, 5, 10. The results obtained are approximately what we would expect by interpolating the corresponding results. For instance, when with two planes, one is a regular network with $k = 2$, and the other is a scale-free one, we obtain a 45% of convergence, which is between 34% (s-f/s-f) and 65% (reg/reg).

For three planes we show the results in Table 3. It is apparent that the connections between agents allowed by the permeability with regular networks yields a significant improvement on the consensus numbers for scale-free networks. Since scale-free networks are frequent in real-world situations, this fact can help to enhance the effectiveness and speed of dissemination of phenomena, for instance for deployment of policies.

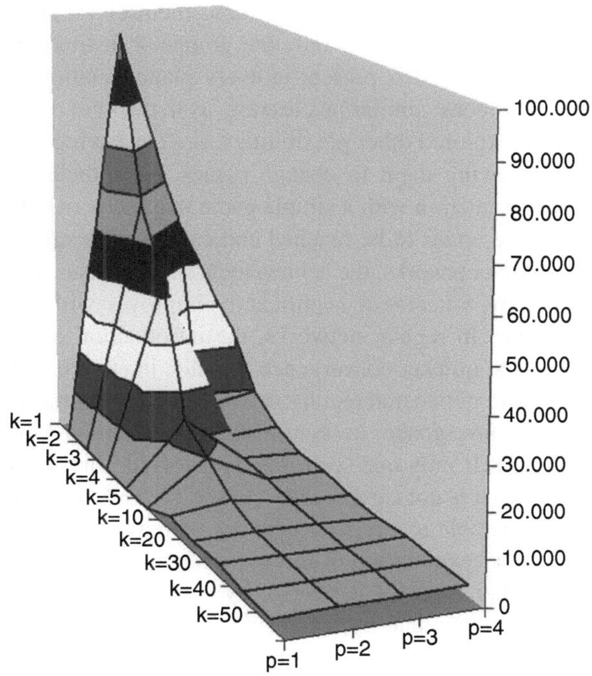

Fig. 2 Average number of meetings to achieve consensus

Table 3 Percentage of consensus achievement, average and standard deviation of the number of meetings for heterogeneous combinations of networks

$k=$	1	2	3	4	5	10	
reg/reg/reg							s-f/s-f/s-f
% Conv	65%	90%	100%	100%	100%	100%	82%
Avg	50,746	6,839	3,845	3,640	3,210	2,530	22,035
St dev	61,658	3,473	2,926	3,938	3,250	1,061	32,113
reg/reg/s-f							
% Cong	72%	88%	98%	98%	95%	98%	
Avg	32,688	26,677	11,817	14,858	10,839	4,946	
St dev	56,992	60,585	25,437	31,245	32,514	7,148	
reg/s-f/s-f							
% Conv	77%	88%	92%	87%	95%	98%	
Avg	26,293	23,798	18,082	20,348	22,354	25,408	
St dev	26,451	44,289	26,630	42,599	46,392	54,061	

8 Conclusions and Future Work

In this paper we have defended the use of explicitly represented multiple relations in multi-agent-based social simulations. Our approach is still quite simplistic in keeping a "first-order" view of the relations: agents only know their own connections, and not of the relations themselves, and the granularity of the society is centred on the agents only (no groups or institutions). On top of that, we assume that every agent is present in every plane (relation), that the several relations are homogeneous: similar in structure, as in the kind of connections between agents. In [3] we explored other possibilities, such as having agents placed in one plane only, and allowing them to change planes. Nevertheless, we show through extensive experimentation with a simple game that the permeability between contexts allows for more nodes to be reached and causes more effective and quicker diffusion. In scale free networks, the achievement of consensus proves impossible with one relation only, whereas it significantly improves with the increase of the number of relations. In regular networks, the introduction of more relations always ensures more and quicker convergence. Finally, in experiments with different types of networks we notice that regular networks of any degree greater than two always induce more convergence to consensus in situations where scale-free networks are involved. If only one scale-free is involved, this convergence is quite faster, while with two it is not significantly slower. Overall, the behaviour of scale-free networks is comparable to a regular network with a small k (2 or 3). Future work will focus on further exploration of this experimental setting, namely by running simulations with other types of networks, by studying alternative policies for the individual agents update, and by increasing the heterogeneity of the networks considered in each run. We will also consider the use of dynamic networks.

Acknowledgements We would like to thank the anonymous reviewers of both revision phases for their feedback which allowed us to significantly improve the paper.

References

1. Albert R, Barabási A-L (2002) Statistical mechanics of complex networks. Rev Mod Phys 74(1):47
2. Amblard F, Ferrand N (1998) Modélisation multi-agents de l'évolution de réseaux sociaux. In: Ferrand N (ed) Actes du Colloque Modèles et Systèmes Multi-agents pour la gestion de l'environment et des territoires, pp 153–168
3. Antunes L, Balsa J, Urbano P, Coelho H (2007) The challenge of context permeability in social simulation. In: Proceedings of fourth European social simulation association conference, Toulouse, France, September 2007
4. Antunes L, Coelho H, Balsa J, Respício A (2006) e*plore v.0: Principia for strategic exploration of social simulation experiments design space. In: Proc. of the first congress on social simulation WCSS 2006, Kyoto, Japan
5. Axelrod R (1997) Advancing the art of simulation in the social sciences. In: Conte R, Hegselmann R, Terna P (eds) Simulating social phenomena. LNEMS, vol 456. Springer, Heidelberg
6. Barabási A-L, Albert R (1999) Emergence of scaling in random networks. Science 286(5439):509–512
7. Brelger R, Carley K, Pattison P (2004) Dynamic social network modeling and analysis: workshop summary and papers. National Academy Press, Washington, DC
8. Hanneman RA, Riddle M (2005) Introduction to social network methods. University of California, Riverside. faculty.ucr.edu/~hanneman

9. Hassan S, Antunes L, Arroyo M (2008) Deepening the demographic mechanisms in a data-driven social simulation of moral values evolution. In: MABS 2008: multi-agent-based simulation. Springer
10. Hassan S, Pavón J, Antunes L, Gilbert N (2010) Injecting data into agent-based simulation. In: Takadama K, Deffuant G, Cioffi-Revilla C (eds) Simulating interacting agents and social phenomena: the second world congress. Agent based social systems (vol 7), Springer, Tokyo (this volume)
11. Masolo C, Vieu L, Bottazzi E, Catenacci C, Ferrario R, Gangemi A, Guarino N (2004) Social roles and their descriptions. In: Dubois D, Welty CA, Williams M-A (eds) Principles of knowledge representation and reasoning: proceedings of the ninth international conference (KR2004). AAAI Press, Menlo Park, CA, pp 267–277
12. Matsuo Y, Hamasaki M, Nakamura Y, Nishimura T, Hasida K, Takeda H, Mori J, Bollegala D, Ishizuka M (2006) Spinning multiple social networks for semantic web. In: Proceedings, the twenty-first national conference on artificial intelligence and the eighteenth innovative applications of artificial intelligence conference, Boston, Massachusetts, USA, 16–20 July 2006. AAAI Press
13. McCarthy J (1993) Notes on formalizing context. In: Proceedings of IJCAI, pp 555–562
14. Seyed-allaei H, Bianconi G, Marsili M (2006) Scale-free networks with an exponent less than two. Phys Rev E 73:046113
15. Urbano P, Decentralised Consensus Games (in Portuguese). PhD thesis, Faculdade de Ciências da Universidade de Lisboa, 2004

A Naturalistic Multi-Agent Model of Word-of-Mouth Dynamics

Samuel Thiriot and Jean-Daniel Kant

Abstract Word-of-mouth is the interpersonal process by which information about a product, and more generally an innovation, diffuses within a social system. Despite of the lack of empirical data on interpersonal communication, several stylized facts are identified and widely accepted. For instance, consumers actively search for information about products. Knowledge about incremental innovations diffuses notably quicker, because part of the knowledge is already available from previous innovations. Hence, existing models applied to word-of-mouth do not reproduce these stylized facts.

We propose an agent-based model in which word-of-mouth is described in a more realistic way. In this model, the representation of individuals' knowledge relies on associative networks. Interactions are described as the motivated communication of the part of beliefs attached to social objects. Simulations illustrate the increased representativeness of the model, including active search for information and diffusion of incremental innovations. These experiments show an important change in the diffusion rate caused by active search for information.

Keywords Agent-based modeling • Social simulation • Information dynamics • Word of mouth • Information search

---

S.T hiriot (✉)
France Télécom R&D, Orange Labs, 4 rue du Clos Courtel, BP 91226, 35512, Cesson-SevignéC edex,Fr ance
and
Computer Science Laboratory (LIP6), Université Pierre et Marie Curie – Paris VI, 4P laceJ ussieu,75005 P aris,Fr ance
e-mail:t hiriot@poleia.lip6.fr

J.-D.K ant
Computer Science Laboratory (LIP6), Université Pierre et Marie Curie – Paris VI, 4P laceJ ussieu,75005 ,P aris,Fr ance
e-mail:J ean-Daniel.Kant@lip6.fr

K. Takadama et al. (eds.), *Simulating Interacting Agents and Social Phenomena: The Second World Congress*, Agent-Based Social Systems 7,
DOI 10.1007/978-4-431-99781-8_7, © Springer 2010

1 Modeling Word-of-Mouth

1.1 Evidence on Word-of-Mouth

In this paper we study word-of-mouth (WOM), that refers to *the process by which knowledge about innovations (new products, practices or ideas) diffuses in a social system.* Nowadays WOM receives increasing attention, not only to enhance viral marketing efficiency, but also to study the diffusion of innovations, the dynamics of opinions, and any other social processes concerned with interpersonal communication.

Data collection for word-of-mouth is hard to set up on a large scale. This explains why so few empirical studies are available (e.g. [3]). Nevertheless, several stylized facts are widely accepted in the field of diffusion of innovations [11], persuasive communication [9], consumer behavior [5] and other domains. Given the lack of quantitative data from the field, these *stylized facts constitute the only basis available to evaluate the plausibility of WOM models.* We summarize here several stylized facts (SF) that are pervasive in literature.

SF 1. Active search for information. It was shown in various fields (impact of advertisement [8], diffusion of innovations [11] and consumer research [5]) that consumer knowledge about a given innovation evolves in several steps. First the consumer is *unaware* of the product. Then, he/she becomes *aware* of the innovation due to some cues received through advertisement or interpersonal communication. Based on these cues and his/her own preferences, he/she can *search actively for information* by retrieving information from his/her friends. Only then can he/she enter in a decision step. When a consumer has enough information, he/she can also *proactively transmit his/her information* to his/her acquaintances. This communicative behavior depends on individual characteristics: a given consumer looks for information, speaks proactively and adopts a product depending to his/her needs, motivations and interests.

SF 2. Partial knowledge transmission about topics When two people meet, they discuss various topics that one or the other have in mind. Obviously they cannot exchange all their knowledge, but rather exchange only their beliefs about the topics of the discussion. Moreover, people often meet and discuss various topics but innovations. During a field study, Carl [3] measured that only 16% of interpersonal communication was about products.

SF 3. Previous knowledge New information received by an individual interacts with previously held knowledge. A piece of information can be misunderstood, so consumers don't fully evaluate the benefits of the innovation [11]. Moreover, a lot of innovations are in fact incremental innovations, which share information with previous products. As pointed out by Rogers [11], incremental innovations diffuse more quickly because part of the knowledge necessary for their adoption is already known from previous innovations.

1.2 Existing Models

Several models were proposed to study information dynamics. The most well known are the epidemics paradigm, the economically-inspired cascade model, and the threshold model.

It is quite intuitive to study information propagation as an *epidemic process* [6]. However in the epidemics viewpoint, *communication is only proactive*: an agent holding information transmits it when it meets someone, but it never actively searches for information as described in SF 1. Moreover, during information transmission all the knowledge held by agents is transmitted when they meet, contradicting SF 2.

In the *stream of economics*, WOM is understood as the transmission of consumers' evaluation of products, rather than their knowledge of the product. Communication is the explicit transmission of payoffs [1, 4] or inference of private cues given observable actions [2]. In these models, agents transmit all their information at each meeting. Hence, these models are not compliant with SF 2 nor 3.

The *threshold model* [7] permits the modeling of social pressure: an agent changes its internal state S for a new one S' if the proportion of its neighbors being in state S exceeds a given threshold. The threshold model enables to describe elegantly the dynamics of opinions, attitudes or culture, for which the multiplicity of sources is a key parameter. But in WOM, one unique interaction with one unique agent is enough to trigger the agent's awareness. Other drawbacks can be underlined: communication with neighbors is systematic (contradicts SF 2) and information is never actively searched (SF1).

1.3 Target

We claim that a more explicit and realistic representation of knowledge can lead to a more powerful model, enabling the compliance with these stylized facts. We previously proposed a model of diffusion of innovations [12] that includes a communication layer describing how agents transmit their beliefs about innovations. In this paper, we show how this communication layer, named CoBAN (for *Communicating on OBjects using Associative Networks*), enables us to model WOM. As described in Sect. 2, the model relies on *associative networks* for knowledge representation, and uses *social objects* to determine the communicative behavior of agents. In Sect. 3, we use simulation to illustrate the new phenomena which can be studied using CoBAN: we model the active search for information, marketing campaigns based on event marketing and the diffusion of incremental innovations.

2 Model

2.1 Structure of Interactions

A social network is used to define the structure of the interactions that take place in the agents population $\mathcal{A}$ (note that notations are summarized in Table 1). At each simulation step, all the interaction links are browsed. For each link, the two agents linked together meet and have the possibility to exchange information. In this paper we use a Watts-Strogatz network [14] which ensures both high cliquiness and low path length. Parameters are set so that $|\mathcal{A}|=1000$ agents and the average path length is 5 (inspired by Milgram's experiment). Given these values, the Watts-Strogatz generator leads to an average degree of 8.

The collective dynamics of the system appears to be sensitive to this structure of interactions. Extended description of that dynamics is out of the frame of this paper. In short, the average path length in the network changes the delay for a piece of knowledge to reach the whole population, while the average degree influences the "tipping point" of the diffusion. Note that, if we use a Watts-Strogatz network for illustration, it should be replaced by a more plausible structure of interactions when the purpose is to describe a given phenomenon. We already proposed better solutions to generate interaction networks from scattered statistics and qualitative observations [13].

Table 1 Table of notations

Notation	Description
$\mathcal{A}$	The set of all agents, representing the artificial population
$\mathcal{C}$	The set of all concepts that may be perceived and understood by agents
$\mathcal{L}$	The set of beliefs (links between concepts) that may exist in the artificial population
$\mathcal{O} \subset \mathcal{C}$	Social objects in the model (topics of interest for several agents within the population)
$\forall a \in \mathcal{A},\ IAN^{a,t}$	Individual Associative Network (belief base) of an agent *a* at time *t'*
$\forall a \in \mathcal{A},\ Sal^{a,t} \subset \mathcal{O}$	The social objects salient for agent *a* at time *t*
$\forall a \in \mathcal{A},\ \forall o \in \mathcal{O},\ Rep_o^{a,t}$	The representation of object *o* held by agent *a* at time *t*
$SR^{o,t}$ with $o \in \mathcal{O}$	A social representation of object *t* held by several agents at time *t*
$wil^{PC,a}: Rep_o^{a,t} \mapsto \mathbb{B}$, $wil^{LFI,a}: Rep_o^{a,t} \mapsto \mathbb{B}$ with $a \in \mathcal{A},\ o \in \mathcal{O}$	Functions describing the communicative behavior of agent *a*. Respectively willingness to speak proactively, and to search information for an object *o* given the representation $R_o^{a,t}$ held at a given time *t* by an agent *a*
$IAN^{\mathcal{X},t_0}$, with $\mathcal{X} \subset \mathcal{A}$	The initial beliefs of a group of agents *x*
TAN	Transmissible Associative Network (message) containing links between concepts

2.2 Knowledge Representation

We assume a network representation of knowledge, as done previously in various streams (e.g. semantic networks, social representations, belief networks, consumers beliefs [10]). Knowledge in the model is defined by a graph $\{\mathcal{C}, \mathcal{L}\}$ in which $\mathcal{C}$ are *concepts* and $\mathcal{L}$ are beliefs formalized by directed links between concepts (that space is restricted by the modeler's hypothesis or data collection, so $\mathcal{L} \subset \mathcal{C}^{2}$). Each agent $a \in \mathcal{A}$ holds its own beliefs base named Individual Associative Network IAN^{a} (for more details, see [12]). For instance in Fig. 1, links contained in $IAN^{a_1,t}$ mean that at time t, agent a_1 trusts that the product iPod is trendy, and that its storage capacity is about 4 Gb.

2.3 Communicative Behavior

As pointed out in SF2, information transmission is about specific topics. The concept of social objects, defined in social psychology as *objects of common interest* [9], is reused to reflect that idea. We define a set of social objects $\mathcal{O} \subset \mathcal{C}$, which are particular concepts of interest for agents. For each social object $o \in \mathcal{O}$, agents can search information for o, proactively communicate on o, and memorize information about o. The representation $Rep_{o}^{a,t}$ of an object $o \in \mathcal{O}$ held by an agent a at time t is defined as the subtree in $IAN^{a,t}$ rooted in o. Messages are named Transmissible Associative Networks (TAN); they contain a representation about one social object that is aimed to be communicated.

To comply with SF1, individuals transmit information depending on their belief state. An agent enters a communicative state, either proactive (PC) or looking for information (LFI), depending to two behavioral functions. *Willingness to look for information* determines if, given the current representation of an object held by the

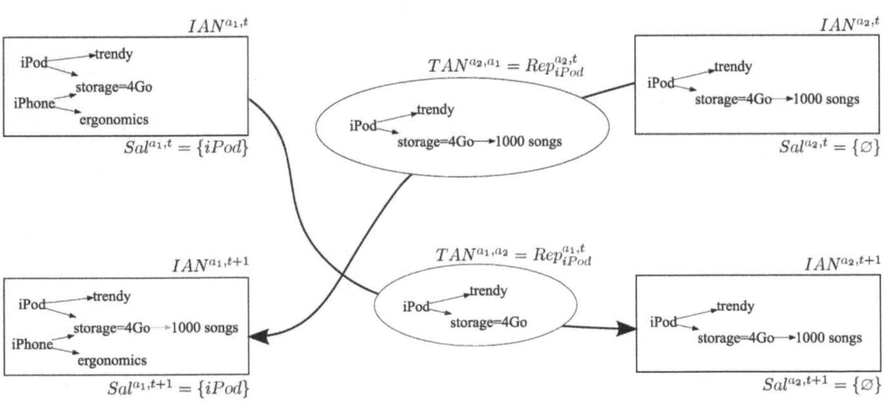

Fig. 1 Example of interaction. Agents a_1 (*left*) and a_2 (*right*) meet a step t

agent, it looks for information about it: $wil^{LFI} : Rep_o^{a,t} \mapsto \mathbb{B}$ ($\mathbb{B}$ is the set of booleans $\{true, false\}$). In the same way, willingness to communicate proactively is determined by $wil^{PC}: Rep_o^{a,t} \mapsto \mathbb{B}$. Both functions depend on the agent's interest in the object (defined by its profile), given beliefs for that object. In states $LFI(o)$ and $PC(o)$, the social object o is said to be *salient* – as it just became a subject of interest – and put in the list of salient objects of the agent noted $Sal^{a,t} \subset \mathcal{O}$. The object becomes unsalient after a timeout fixed to four steps.

The interaction protocol between two agents $a_1, a_2 \in \mathcal{A}$ includes three steps. An example of interaction is depicted in Fig. 1. (1) *Determination of discussion topics* based on interests of each agent: $O^{a_1,a_2,t} = \{S^{a_1,t} \cup S^{a_2,t}\}$. If $O^{a_1,a_2,t} = \{\varnothing\}$ agents do not discuss anything (consistent with SF2). (2) For each object $o \in O^{a_1,a_2,t}$, each agent retrieves from its $IAN^{a,t}$ the representation $Rep_o^{a,t}$. That representation is transmitted to the other agent as a TAN^{a_1,a_2}. (3) Beliefs of each agent are updated with the representation of the other. If the agent's representation (e.g. a_1 in Fig. 1) did not include some links of this TAN, then the agent adds these new links in its IAN. Otherwise, no changes are made[1] (e.g. agent a_2 in Fig. 1).

3 Simulation

3.1 Dynamics with Active Searches

In order to illustrate the qualitative benefits of the model, we propose here a stylized diffusion of information. Let us suppose that a firm tries to diffuse information for an innovation, for instance iPod (chosen only as an illustration, without ambition of realism). Advertisement campaigns can only provide some cues to make consumers aware of the product, and maybe raise their interest and lead them to search information. Here the communication campaign transmits two cues, namely the storage capacity of the device and the fact that iPod is perceived as trendy. The TAN representation TAN_{Ad}^{iPod} of that advertisement is depicted in Fig. 2. Each agent has an exposure probability $exp^{Ad} = 0.05$ to receive that message at each step. However, one must have previous knowledge about storage size in order to work out the quantity of songs which can be stored in the device. Some of the consumers, named *experts* (0.5% of the population), are already aware of that consequence, so their IAN is initialized with the chains IAN^{exp,t_0} (cf. Fig. 2) (Table 2).

Communicative behavior is defined as follows. A proportion $p^{curious} = 0.1$ of agents are *curious*: when they believe that an object is trendy, they look for more information about that object. Formally it is expressed by $wil^{LFI}(Rep_{iPod}^{a,t}) = (trendy \in Rep_{iPod}^{a,t})$. In the same way, a proportion $p^{promoters} = 0.2$ of agents are made *promoters*: if

[1] Belief revision is simplified in this paper for shake of clarity. The complete model, described in [12], also manages belief revision with contradictions, based on both beliefs' and emitters' credibilities.

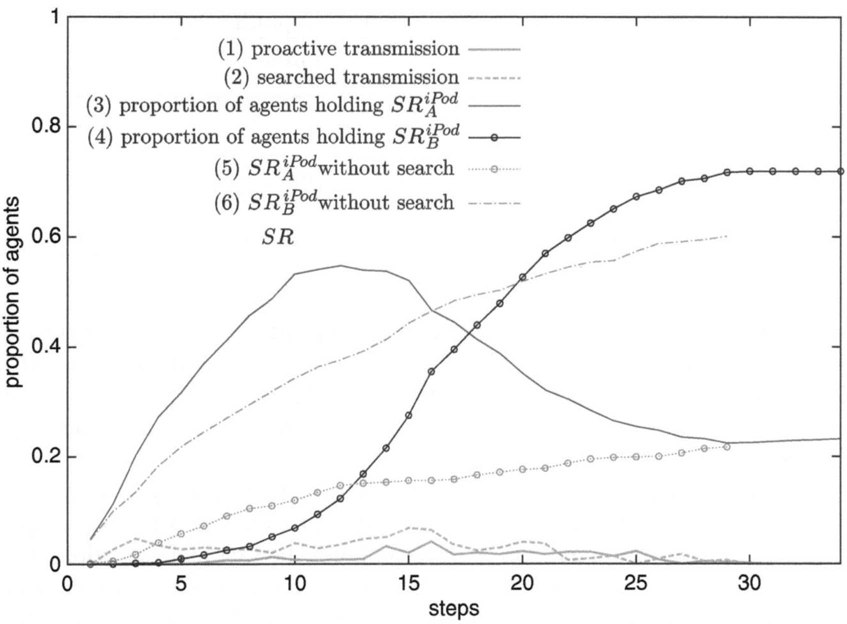

Fig. 2 Word-of-mouth dynamics for a social object iPod with active search for information. Two social representations appear in the population (SR_A^{Ipod} and SR_B^{Ipod})

Table 2 Social representations appearing in the experiment depicted in Fig. 2

TAN^{iPod}	IAN^{exp,t_0}	SR_A^{iPod}	SR_B^{iPod}
Ipod → trendy Ipod → 4Go	4Go → 1000 songs	Ipod → trendy Ipod → 4Go	Ipod → trendy Ipod → 4Go → 1000 songs
(parameter)	(parameter)	(indicator)	(indicator)

they trust that an iPod can store as many as a thousand songs, they speak proactively about it. The problematic of the firm is to understand what percentage of the population will receive enough knowledge for them to understand the benefits of the innovation. In other words, the question is the efficiency of interpersonal communication to retrieve information held by some agents, given advertisement which makes some agents look for more information.

Figure 2 represents the diffusion of information about the iPod. The first two curves (1 and 2) represent the amount of interpersonal communication, respectively due to proactive communication and information searches. During simulation, *social representations* – that is, representations of a social object shared by several agents [9] – appear in the system. These social representations, later noted SR , provide a useful indicator to monitor the diffusion of knowledge in the population. Curves 3 and 4 display the diffusion of social representations, whose content is

Table 3 Social representations appearing in the experiment depicted in Fig. 3

TAN^{Event}	SR_A^{Event}	SR_B^{Event}
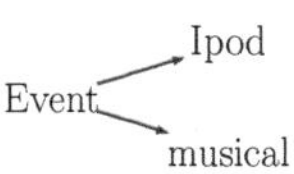	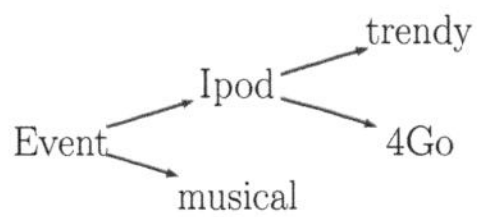	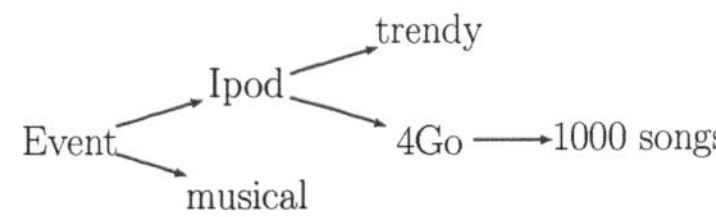
(parameter)	(indicator)	(indicator)

detailed on Table 3. During the advertisement campaign, several social representations appear. The first one, SR_A^{iPod}, drawn as curve 3, corresponds to agents which only received information from advertisement. The more detailed representation SR_B^{iPod} is held by agents which retrieved both advertisement and expert knowledge.

We observe that active searches allow agents in the first category to retrieve very efficiently the more detailed representation, even with as few as 0.1% expert agents in the population. Indeed, as soon as an agent holds the first representation, it will either look for information and retrieve the detailed representation, or meet a promoter which will probably inform it. This will cause a switch from SR_A^{iPod} to SR_B^{iPod}, and then a drop in the proportion of SR_A^{iPod}, as shown by curve 3 in Fig. 2. Curves 1 and 2 show that the shift from simple to detailed knowledge is mainly due to active search for information and secondly by proactive communication.

As a reference we made the same simulation by setting $p^{curious} = 0$, so the model becomes purely epidemic. The resulting curves (5, 6) show simulation results with all parameters equal. Only the less detailed representation SR_A^{iPod} is transmitted (curve 5) while the more detailed SR_B^{iPod} (curve 6) remains the minority. This is because agents lack the active search skills to switch from elementary SR_A^{iPod} to elaborated SR_B^{iPod}. That experiment suggests that taking into account the consumers' ability to search for information has a huge influence on the diffusion of knowledge.

3.2 *Using an Event to Facilitate Diffusion of Information*

At the end of the advertisement campaign, the institution observes that only a part of the target population received the detailed information. Launching another communication campaign would not raise interest of consumers one more time. A solution, used in the field, is to organize a popular event, which will create word-of-mouth, in order to make people exchange information about the product. An event is organized, with property "musical program". The proportion of promoters of such an event is set to 0.2. Communication about the event is created so people strongly associate *event* with the *iPod*; the information TAN^{Event} disseminated in the population is depicted in Fig. 3. Arbitrarily only 20 random agents out of 1,000 are made aware of that representation at step 35 (Table 4).

Simulation results (Fig. 3) show the impact of that campaign: while the information about the event diffuses (curves 6 and 7), the proportion of agents holding the

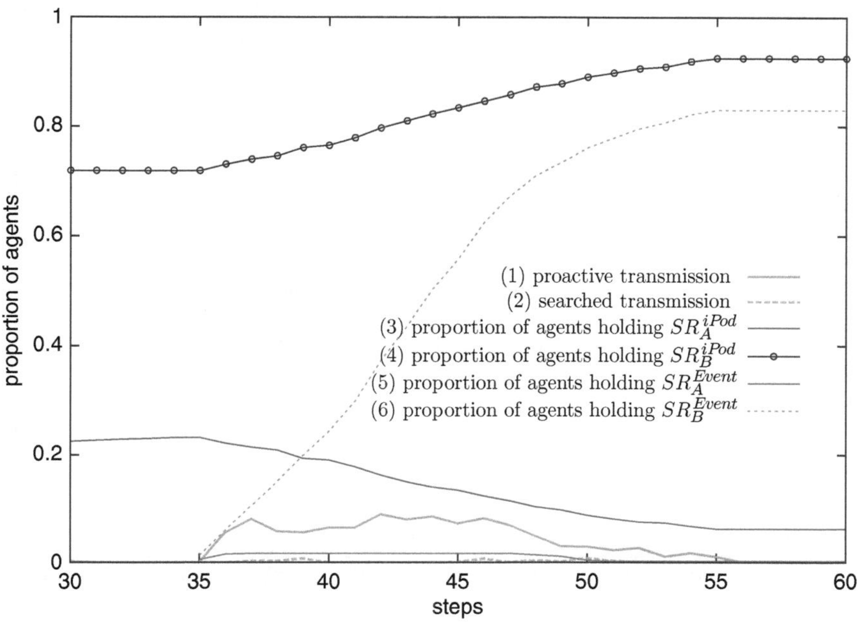

Fig. 3 Diffusion of information about an event related to iPod permits enhanced knowledge about iPod

Table 4 Social representations appearing in the experiment depicted in Fig. 4

TAN^{iPhone}	SR_A^{iPhone}	SR_B^{iPhone}
Iphone → 4Go Iphone → breakable	Iphone → 4Go Iphone → breakable	Iphone → 4Go → 1000 songs Iphone → breakable
(parameter)	(indicator)	(indicator)

detailed representation of iPod SR_B^{iPod} (curve 3) also shifts by 10%. These simulations illustrate that with associative networks, modeling the diffusion of knowledge about related social objects becomes trivial.

3.3 *Diffusion of Related Products*

At time $t = 60$, the firm launches an incremental innovation, namely the iPhone. The content of advertisement TAN^{iPhone} is provided in Fig. 4. Note that the iPhone shares one common attribute with the iPod: the storage size. As in the previous diffusion of the iPod, understanding its storage size requires some knowledge. However, agents in the population already learned that information during the communication

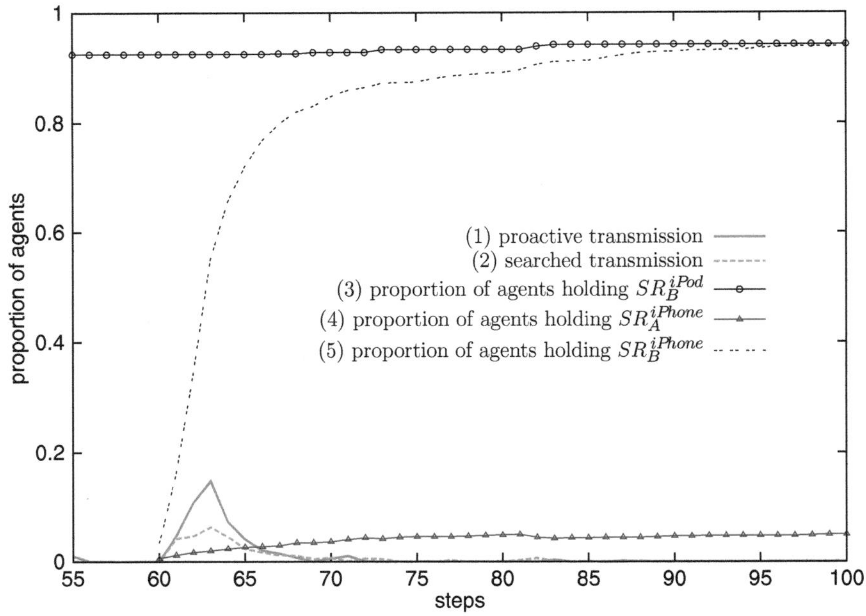

Fig. 4 Diffusion rate of knowledge about a related innovation, here the iPhone, is increased by previous knowledge

about the iPod. The diffusion curve in Fig. 4 shows a nearly instantaneous diffusion of that information; almost no agent remains with only the information received by advertisement. Most agents were already aware of detailed information of the iPod (curve 3) and are immediately able to understand that the storage capacity is valuable. That simulation, even if highly stylized to simplify our demonstration, clearly shows that CoBAN complies with SF 3.

4 Discussion

In this paper we proposed a model of information diffusion, named CoBAN, in which beliefs about social objects are represented by associative networks. We showed that using social objects avoids unrealistic systematic exchange, compliant with SF 2. Associative networks allow the satisfaction of the "partial knowledge exchange" property, by defining communication as the emission of the representation of an object rather than transmission of the whole belief base. Associative networks also allow the modeling of active search for information (SF 1). The impact of active search on dynamics appears to be strong. The CoBAN model is also closer to consumer/adopter behavior models [5, 8, 11], and complies with the actual marketing process of creating cues to raise interest. Simulations prove that the model can be used to study the diffusion of incremental innovations or event marketing (SF3).

CoBAN is a more plausible model, anchored on indisputable stylized facts. Its dynamics has now to account for quantitative data taken from the real world. However, it is very difficult to obtain such data, that is to measure the amount and the nature of information exchanged in face-to-face communication. Therefore we plan in the very next future to collect real data on the behavior level of product adoption, and to assess CoBAN using this data.

Acknowledgement Part of this work was funded by research grant CIFRE 993/2005 from the French National Association for Research and Technology (ANRT). Support was also provided by France Télécom R&D – Orange Labs.

References

1. Bala V, Goyal S (1998) Learning from neighbours. Rev Econ Stud 65(3):595–621
2. Bikhchandani S, Hirshleifer D, Welch I (1992) A theory of fads, fashion, custom, and cultural change as informational cascades. J Polit Econ 100(5):992–1026
3. Carl WJ (2006) What's all the buzz about? Everyday communication and the relational basis of word-of-mouth and buzz marketing practices. Manag Commun Q 19(4):601–634
4. Ellison G, Fudenberg D (1995) Word-of-mouth communication and social learning. Q J Econ 110(1):93–125. doi:10.2307/2118512
5. Engel JF, Blackwell RD, Miniard PW (1995) Consumer behaviour, 9th edn. The Dryden Press, Orlando
6. Goffman W, Newill V (1964) Generalization of epidemic theory: an application to the transmission of ideas. Nature 204(4955):225–228
7. Granovetter M (1978) Threshold models of collective behavior. Am J Soc 83:1360–1380
8. McGuire WJ (1989) Public communication campaigns, 2nd edn. Chap. Theoretical foundations of campaigns. Sage Publications, Newbury Park, pp 43–65
9. Moscovici S (1998) Psychologie Sociale, 7th edn. Presses Universitaires de France, Paris
10. Reynolds TJ, Gutman J (1988) Laddering theory, method, analysis, and interpretation. J Advert Res 28:11–31
11. Rogers EM (2003) Diffusion of innovations, 5th edn. Free Press, New York
12. Thiriot S, Kant JD (2007) Representing knowledge as associative networks to simulate diffusion of innovations. In: Amblard F (ed) Proceedings of ESSA'07, the 4th conference of the European social simulation association, Toulouse, France, 10th–14th September 2007, pp 193–204
13. Thiriot S, Kant JD (2008) Generate country-scale networks of interaction from scattered statistics. In: The fifth conference of the European social simulation association, Brescia, Italy
14. Watts DJ, Strogatz SH (1998) Collective dynamics of 'small-world' networks. Nature 393(6684):440–442

Part II
Economy, Market and Organization

Introducing Preference Heterogeneity into a Monocentric Urban Model: AnA gent-BasedL andM arketM odel

Tatiana Filatova, Dawn C. Parker, and Anne van der Veen

Abstract This paper presents an agent-based urban land market model. We first replace the centralized price determination mechanism of the monocentric urban market model with a series of bilateral trades distributed in space and time. We then run the model for agents with heterogeneous preferences for location. Model output is analyzed using a series of macro-scale economic and landscape pattern measures, including land rent gradients estimated using simple regression. We demonstrate that heterogeneity in preference for proximity alone is sufficient to generate urban expansion and that information on agent heterogeneity is needed to fully explain land rent variation over space. Our agent-based land market model serves as a computational laboratory that may improve our understanding of the processes generating patterns observed in real-world data.

Keywords Agent-based markets • Land use • Preferences heterogeneity

---

T.Fila tova (✉)
Centre for Studies in Technology and Sustainable Development, Faculty of Management andG overnance,U niversityof T wente, Drienerlolaan5, 7522 N B,E nschede,T heN etherlands
e-mail:T .Filatova@utwente.nl

D.C.P arker
School of Planning, University of Waterloo,
200U niversityA ve.W est,W aterloo,O nt.,N 2L3G 1,C anada
e-mail:dc parker@uwaterloo.ca

A. van der Veen
Department of Water Engineering and Management, University of Twente,
Drienerlolaan 5, 7522 NB, Enschede, The Netherlands
and
International Institute for Geo-Information Science and Earth Observation, University of Twente,
Hengelosestraat, 99,P.O.B ox6, 7500 A A,Ens chede,The N etherlands
e-mail:V een@itc.nl

K. Takadama et al. (eds.), *Simulating Interacting Agents and Social Phenomena: The Second World Congress*, Agent-Based Social Systems 7,
DOI 10.1007/978-4-431-99781-8_8, © Springer 2010

1 Introduction

Spatial forms of cities and urban land prices are the results of land allocation between competitive users via land markets. Land market models in urban economics, as many other economic models, often assume a single representative agent [1, 2]. This paper presents an agent-based model of an urban land market in which agents exhibit heterogeneous preferences for proximity to the urban center. We compare macro scale economic and spatial measures arising from both homogeneous and heterogeneous agents interacting in a land market. We show that by providing the opportunity to track characteristics of agents, the spatial goods being exchanged, and associated land transaction data, our agent-based land market serves as a computational laboratory for exploring micro-macro linkages in urban land markets (particularly, links between individual preferences, emerging land prices and urban patterns). In fact, ABM provides a methodological platform for direct modeling of the land market, as does urban economics, and statistical analysis of the land rent function, as does spatial econometrics. Through modeling, it reveals potential processes that may stand behind the aggregated indices analyzed by econometrics.

We underline the importance of building from existing theoretical and empirical work done in spatial, urban and environmental economics in constructing our ABM of land use with an endogenous market mechanism. Many traditional models of urban land markets find their roots in the monocentric urban model of W. Alonso [1]. According to his bid-rent theory, households choose locations at a certain distance from the central business district (CBD[1]) by maximizing utility from the joint consumption of a spatial good (a land lot or house) and a composite good (all other goods) under their budget constraint. The outcome of the bid-rent model is a set of rent gradients (i.e., land prices at different distances from the CBD). The model predicts that the land rent gradient is decreasing with distance and land prices for equidistant locations are the same.

As is typical in economics, certain restrictive assumptions are made to solve for equilibrium conditions in traditional urban economics models. In general these restrictive economic assumptions can contradict real world phenomena and have raised substantial criticisms. These assumptions (each of which has a representative example in urban economics) fall into four general areas [3]: limitations of *the representative agent* approach [4]; limitations of assumptions of economic *rationality* [5]; absence of direct *interactions among agents* [6]; and absence of analysis of *out-of-equilibrium dynamics* [5, 7, 8]. As discussed by many agent-based computational economics scholars, ABM may serve as a tool to relax one or several of these assumptions to shift to more realistic models. For the purposes of this paper, we introduce heterogeneous agents and replace an equilibrium centralized price determination by distributed bilateral trading.

---

[1]CBD is assumed to be exogenously given. For future work it might be interesting to explore model dynamics with endogenous formation of CBD and suburban centers.

Applications of ABMs to land use (ABM/LUCC) are quite diverse [9]. To date, the majority of efforts of the ABM community to integrate markets into ABM/LUCC have been focused on agricultural applications [10–12], which differ from urban land markets [13]. Several models study the effects of hypothetical urban land markets, but with primary emphasis on the demand side. The SOME and SLUCE models allow agents to choose the parcel that maximizes their utility without competition from other sellers and assuming that the locating agent will outbid the current use [14]. The MADCM model provides a welfare analysis of the simulated urban land market [15] with the focus on the demand side. Our model moves beyond previous work in several aspects. Both the demand and supply sides are represented in detail, facilitating model experiments focused on the drivers of each.[2] The process of locating trading partners in space, forming bid and ask prices, and resolving trades is also modeled explicitly. The primary aim of this paper is to investigate *how aggregated land patterns and rent gradients change in the monocentric urban model if agents with homogeneous preferences for proximity are replaced by heterogeneous ones*. In the coming sections we describe the structure of our model and discuss the simulation results.

2 The Model

Our *Agent-based Land MArket* (*ALMA*) model explicitly represents micro-scale interactions between buyers and sellers of spatial goods. In line with the assumptions of the monocentric model, the ALMA model assumes that sellers (i.e. owners of agricultural land) offer land at the same fixed price equal to agricultural opportunity costs and that each spatial good is differentiated by distance from the CBD (or its inverse measure – proximity *P*[3]), while environmental amenities (*A*) are assumed to be distributed uniformly in the city. Buyers (i.e., households) search for a location that maximizes their utility

$$U = A^{\alpha} \cdot P^{\beta} \tag{1}$$

Here α and β are individual preferences for green amenities and proximity, correspondingly, and utility, as usual in micro economics, is a mathematical representation of preferences. The choice of land to buy is constrained by an individual budget (*Y*) net of distant-dependent (*D*) transport costs (*T*):

$$Y_{net} = Y - T \cdot D \tag{2}$$

The rationality of agents is bounded by the fact that they do not search for the maximum throughout the whole landscape but rather search for the local maximum

[2] In this particular paper we replicate the monocentric urban model that assumes that sellers are agricultural land owners and that their ask price is the same for every cell. However, the code of our program integrates the possibility to model the formation of ask prices for households and agricultural sellers.

[3] Proximity is defined as $P=D_{max}+1-D$, where D is distance of a cell to the CBD.

among N randomly chosen cells. We impose this assumption since the search for a house in reality is very costly (time-wise and money-wise), meaning that a global optimum is not likely to be located in real-world housing markets.

After defining the spatial good that gives maximum utility, a buyer forms her bid price. The demand function for land of a single buyer, i.e. her willingness to pay for land (WTP), depends on her utility (U), individual income (Y) and prices of all other goods (the influence of which is expressed by a constant b)[4]:

$$WTP = \frac{Y \cdot U^n}{b^n + U^n} \tag{3}$$

Here, Y and U are calculated according to (1) and (2) respectively, and b is a constant. The WTP function is monotonically increasing approaching Y as $U \rightarrow \infty$, meaning that individual WTP increases with utility but does not exceed individual budget. The value of parameter b controls the steepness of the function. As $b \rightarrow \infty$ the function in (3) becomes flatter. We can think of b as a proxy of the affordability of all other goods to reflect their relative influence on the WTP for housing. The WTP function in (3) exhibits the main properties of the neoclassical demand function. Specifically, demand for land grows with income and utility from land consumption but decreases with distance to the CBD. Also, demand for housing decreases as b increases. Since the prices for non-housing goods increase while income remains constant, the share of budget for housing decreases because of the additional expenses for non-housing goods.

In the ALMA model it is assumed that the WTP and final bid price of a buyer might differ. The reason is that buyers (and sellers) try to maximize their gains from trade, i.e. difference between their WTP (willingness to accept – WTA) and transaction price. In this paper, we assume that buyers form their bid price equal to their true WTP ($P_{bid} = WTP$). Other pricing strategies and their effects on the division of gains from trade are discussed elsewhere [3].Then buyers submit their offer-bid to the sellers. Sellers choose the highest offer-bid and if it is above their WTA, then transactions take place. If not then both buyer and seller participate in the land market in the next time step. The final transaction price is an arithmetic average of the ask price and the highest bid price. Figure 1 shows the logic of the trading mechanism, i.e. one time step in the model.[5] The model stops running when no more transactions occur, i.e. all the submitted bids are lower than sellers' WTA.

Obviously buyers and sellers are not the only agents participating in a real-world land market. Urban developers and real estate agents influence both spatial patterns and land price formation. We discuss their roles and ways to integrate them in the ALMA model elsewhere [3, 16]. In this paper we keep the model as simple as possible to give an analysis of the effects of *agents' preference heterogeneity* on the

[4]The justification and properties of this demand function are discussed in detail in [3].

[5]For extended description of the event sequencing see [3].

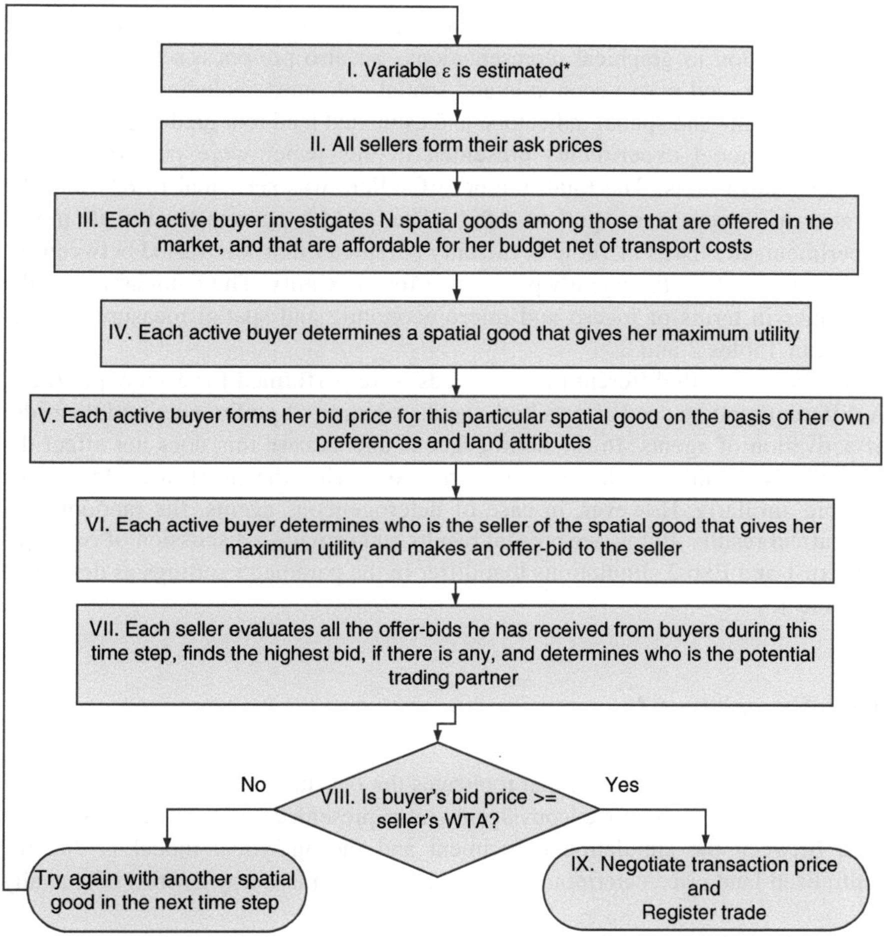

Fig. 1 Conceptual algorithm of trade (*this variable ε is explained in detail in [3])

urban pattern and land prices. Additional examples of agent-based modeling that integrate real estate agent and developers are Hawksworth et al. [17] and Robinson and Brown [18], correspondingly.

3 Simulation Experiments

The model simulations produce spatially explicit rent gradients and land patterns. Experiments varying different parameters such as transport costs, bidding strategy, the level of environmental amenities and others have been performed with the model [3, 13]. In this paper we investigate how changes in buyers' and sellers' preferences, particularly a shift from homogeneous to heterogeneous preferences for proximity to the urban center, affect economic indicators and the spatial morphology of the city. In addition to graphical representations, we also present a set of metrics to analyze micro and macro economic and spatial outcomes, including welfare measures, economic and spatial indicators, and estimated land rent gradients.[6]

All the model experiments presented in this paper were performed on a 29×29 grid of cells. The total number of sellers was set equal to 841 and the number of buyers was equal to 1,000. The ALMA parameters for all model experiments are listed in Table 1; the only parameter that was varied between the two experiments is the agent's preference for proximity. The comparison of the outcomes in terms of macro and micro economic and spatial measures is presented in Tables 2 and 3.

Thirty runs with different random seeds were performed for each experiment. A different random seed affects both the distribution of preferences and the order of activation of agents. In the homogeneous agents case this does not affect the macro metrics: although the order of activation varies, the agents are all the same and bid similarly. However, in case of heterogeneous agents, the random seed does affect results. Below we present results and provide a discussion of two typical Exp 1 and Exp 2 simulations that differ in the parameter settings as described in Table 1.

3.1 Experiment 1

We begin with an experiment that replicates the benchmark case of a monocentric Alonso model with homogeneous agents (also presented in [3]). The main difference between the simulation experiment and the analytical model is that the centralized land price determination mechanism is replaced by a series of spatially

Table 1 Values of parameters in the simulation experiments

Symbol	Meaning	Exp 1	Exp 2
Y	Individual budget	800	800
A	Level of green amenities	1	1
b	Constant in (1)	70	70
N_{cells}	Number of spatial goods (lots) in the city	841	841
P_{ask}	Ask price of a seller of agricultural land	250	250
TCU	Transport costs per unit of distance	1	1
β	Individual preference for the proximity to the CBD	0.85	uniform distribution [0.7;1]
–	Mean preference in the traders population	0.85	0.85

[6] An equation that quantitatively characterizes the transaction price at a given distance from the city center, estimated using linear regression analysis. The land gradient is a typical characteristic of urban spatial structure analyzed both theoretically and empirically [2].

Table 2 Economic and spatial metric outcomes of the ALMA experiments

Parameter	Exp 1	Exp 2
Individual utility:		
Mean	65.48	65.69
St. dev.	12.56	13.07
Aggregate utility	30447.22	33423.41
Buyers' bid price:		
Mean	363.73	364.33
St. dev.	73.92	76.67
Urban transaction price:		
Mean	306.86	307.16
St. dev.	36.96	38.33
Average surplus:		
Buyers'	50%	50%
Sellers'	50%	50%
Total property value:		
Mean	142690.2	156284.2
St. dev.	0	2134.49
City size (urban population):		
Mean	465	508.8
St. dev.	0	8.17
Distance at which city border stops:		
Mean	12.08	12.97
St. dev.	0	0.17

Table 3 Linear regression estimates of rent gradient functions (Transaction price is the dependent variable)

		Exp 2 (1 and 2)	
Parameter	Exp 1	1 Independent variable (D)	2 Independent variables (D and β)
Number of observations	465	507	507
R^2*:	*0.9905*	*0.9560*	*0.9832*
Intercept:			
Estimate	410.76	414.16	482.32
St error	0.501	1.084	2.477
t-Value	819.68	381.99	194.71
Distance to CBD:			
Estimate	−12.81	−12.64	−11.98
St error	0.058	0.121	0.078
t-Value	−219.94	−104.70	−153.32
Buyers' preference for proximity:			
Estimate	–	–	−93.50
St error	–	–	3.271
t-Value	–	–	−28.58

*95% Confidence intervel

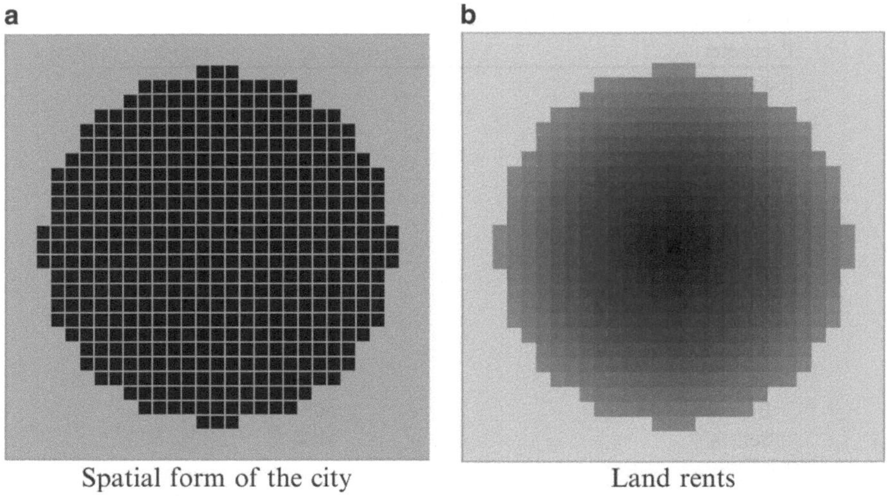

Fig. 2 Exp 1, Replication of the Alonso model (with homogeneous preferences for commuting)

distributed bilateral trades. The results are presented in Table 2. The spatial form of the city and urban land rent gradient are presented in Fig. 2a, b respectively.

The light area in Fig. 2a represents agriculture and the black–the urban area. The intensity of grey color in Fig. 2b symbolizes the value of land: the darker the color, the higher the land price. As in the benchmark case of a theoretical monocentric urban model, the land rent gradient is decreasing with distance. The urban land price is equal for cells equidistant from the CBD. The city expansion stops at the location where the bid price of a buyer falls below the agricultural rent. The lightest-grey area in Fig. 2b shows the beginning of agricultural area (urban-rural fringe) and symbolizes the city border. Note that not all of the buyers in the system ultimately purchase properties (only 465 of the 841 buyers engage in transactions). The parameter settings for Exp 1, then, replicate an open city model, where buyers are assumed to have the opportunity to purchase property in another location, if their bid price for available properties in this region is below the sellers' ask prices of the current land owners.

By applying a simple regression analysis to the model-generated data, we estimated a land-price gradient[7] (Table 3 and Fig. 3). Figure 3 shows a down-slopping demand function for land as a function of distance in the simulated urban zone. The horizontal axis shows the distance from the CBD in spatial units (a spatial unit can be interpreted as 1 km or 1 mile, although in our generalized analysis we do not refer to any specific unit here). The vertical axis shows the urban land price in monetary units (such as dollars or euro).

[7]The results of the linear regression model showed the best fit. The R^2 values for linear, log-log, semi-log and inverse semi-log functional forms were 0.9923, 0.8166, 0.9738 and 0.8647 respectively.

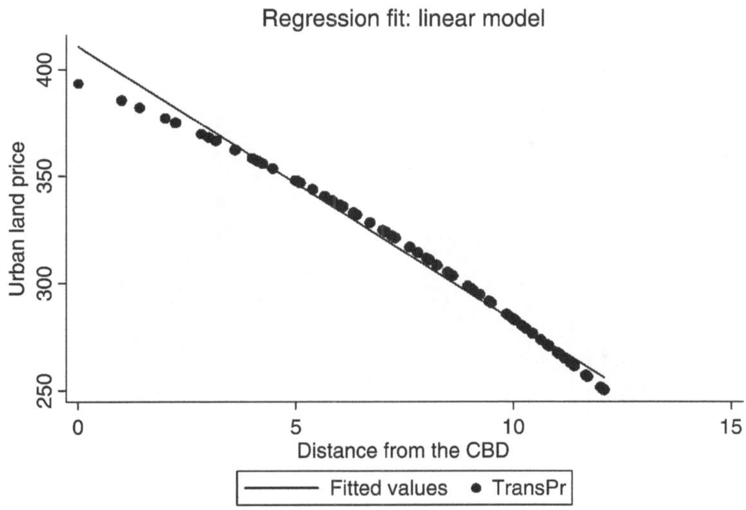

Fig. 3 Land rent gradients for Exp 1, linear regression fit of the model-generated data. *TransPr* – actual land transaction prices, *Fitted value* – estimated land rent gradient

3.2 Experiment 2

The setup in this experiment is almost the same as in Exp 1 (see Table 1) except for the fact that agents are heterogeneous with respect to preference for proximity. We assume that agents have different tolerances for commuting; i.e. β from the utility function follows a uniform distribution in the range [0.7; 1] with mean equal to 0.85.[8] For a linear city (i.e.1D) a theory of monocentric land use equilibrium for households with taste heterogeneity was developed [19]. To our knowledge, an analytical calculation of equilibrium land prices is not possible for this heterogeneous agent case in the two-dimensional landscape.

The first difference from Exp 1 manifests itself in the spatial morphology of the city, as can be seen from comparison of Figs. 2a and 4a. The city border has expanded and the urban population has increased (from 465 to 508.8) as can be seen in Table 2. These differences between two experiments are statistically significant at a 95% confidence level as shown with a *t*-test (the null hypothesis of no difference in means is rejected at $p<0.001$). Thus, if agents have different tolerance levels for commuting and are not constrained by the remoteness of the location (e.g., use private cars instead of public transport) then this is already enough to cause the urban area to sprawl even if green amenities are distributed homogeneously across the city. Essentially, heterogeneity in individual preferences for proximity may be a contributor to urban sprawl.

[8]We also ran the model with the normal distribution of preferences. The results were qualitatively similar.

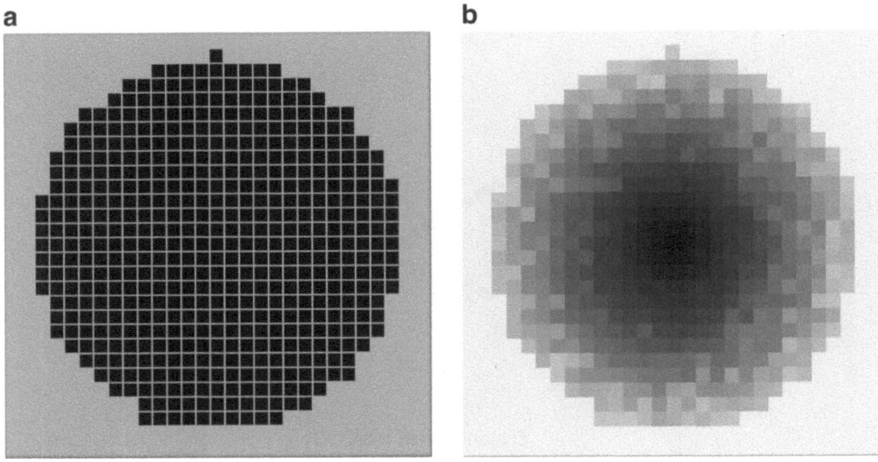

Fig. 4 Exp 2, Monocentric urban model with heterogeneous agents (with respect to commuting preferences)

Additionally, the city no longer expands uniformly in all directions (compare the south, north, west and east borders of the city on Fig. 4a). Since people have different preferences for proximity, there are just a few individuals tolerant enough to commuting to locate at the most distant edges of the city, such as the person in the northern-most cell of the city (Fig. 4a).

Land rents (Fig. 4b) are still decreasing with distance as in Exp 1 (Fig. 2b). However, in Exp 2 the prices of the cells at the same distance from the CBD are no longer equal because of preference heterogeneity. Thus, neither the spatial form nor land prices are symmetric in the city with heterogeneous agents. The average transaction price is slightly higher in the city with heterogeneous preferences for proximity to CBD than in the homogeneous case. This result is not statistically different between Exp 1 and Exp 2, but that lack of significance is easily explained. The change in model parameters has shifted the land rent gradient, rather than primarily affecting average rents. On the one hand, agents with higher than average preferences for proximity, i.e. $\beta > 0.85$, bid more for the urban land that is closer to the CBD than an average agent, i.e. $\beta = 0.85$ as in Exp 1. On the other hand, agents with lower than average preferences for proximity, i.e. $\beta < 0.85$, bid more for the remote spatial good than an average agent from Exp 1 would do, because the former are more tolerant to commuting. In Exp 2 the average β of the agents who actually settled in the city is 0.79, meaning that agents more tolerant to commuting out bid agents with strong preferences for proximity to CBD. Thus, the agents with characteristics at the tails of the distribution drive model outcomes in the heterogeneous case.

The total property value is about 9.5% higher in Exp 2 than in Exp 1 (see Table 2). This result is significant at a 95% confidence level as demonstrated by a *t*-test. This total value is higher for two reasons: first, the agents with higher com-

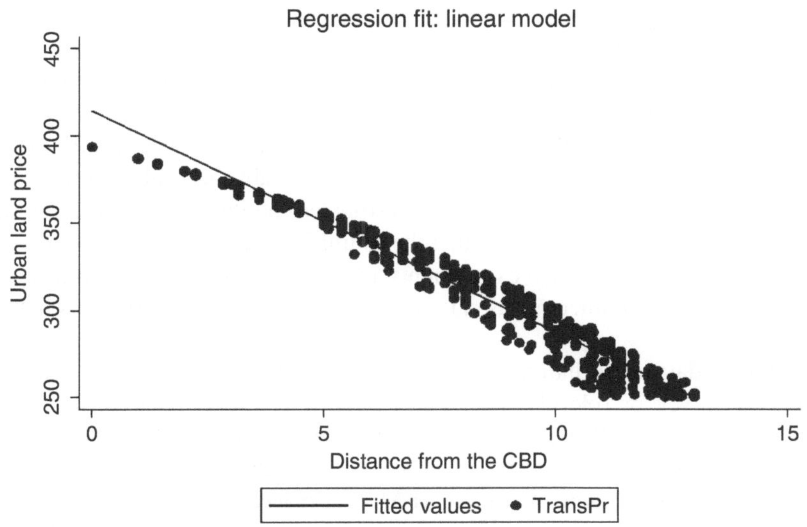

Fig. 5 Land rent gradients for Exp 2-1. Linear regression fit of the computer generated data. *TransPr* – actual land transaction prices, *Fitted value* – estimated land rent gradient

muting tolerances and therefore higher WTP are winning the bids for the properties farther from city center, meaning higher prices for these properties, and second, more cells are converted to urban than before.

The estimated land rent gradient of the computer-generated data from Exp 2 is presented in Fig. 5 and in Table 3. From Fig. 3 it can be seen that transaction data are almost on the regression line. In contrast, data on transaction land price from Exp 2 are more dispersed, but still downward sloping as in Alonso's bid rent theory. The dispersion (essentially, distance-dependent heteroskedasticity) arises from the preference heterogeneity. Standard econometric theory also tells us that this rent gradient estimate is biased due to the omitted variable of preference heterogeneity. This simple modeling exercise illustrates that observed variation in real-world transaction prices may arise from non-spatially-uniformly distributed unobserved agent-level characteristics rather than from unbiased random error. Thus, rent gradient estimates that do not control for agent-level heterogeneity are likely to be systematically biased.

The explained variation (R^2) from Exp 1 is higher in Exp 1 than Exp 2-1 (0.9905 vs. 0.9560, Table 3). The settings for Exp 1 are very abstract, especially in its assumption of homogeneous preferences for location. If everybody in the city behaves as a *representative agent* does then *land prices can be fully explained only by the characteristics of the spatial environment*, such as distance. In practice, this is the only information usually available for hedonic price estimation. However, in reality agents' preferences for the spatial good vary. Therefore, only a portion of the percent of variation of the land price is explained by land characteristics.

The ABM environment allows us to link information about agents' preferences to transaction data and analyze it in the regression analyses. We re-ran the regression model using data from Exp 2, including agents' preferences for proximity as a second independent variable (see Table 3, column 4). As expected, the explained variation in land prices increases (i.e. R^2 is 0.9832 instead of 0.9560). Further, the estimated coefficient on distance to CBD declines in absolute value from 12.64 to 11.98, indicating that the first model overestimated the influence of transportation costs on hedonic land values. Note that the estimated coefficients on distance differ substantially between the two experiments and are each significant, demonstrating that the change in parameters has caused a shift in the land rent gradients between the experiments.[9]

Certainly, data about individual preferences is not easily acquired. Usually this requires a survey or role-playing games. Such micro level data can be used to construct empirical ABMs. ABMs fed with empirical data at the micro decision level [14, 20] provide nice examples of analysis of both the *spatial* and *agent-level drivers of* land-use change. Thus, an ABM land market supported by micro-level data on preferences for location can serve as a computational laboratory in which we have a full understanding of the agent-level and spatial factors that influence *bid prices, ask prices, and realized transactions.*

4 Conclusions and Discussions

In this paper, we have presented an agent-based land market model and analyzed the macro outcomes of simulations for the case of homogeneous and heterogeneous agents' preferences. For this purpose only the variation in one parameter (i.e. agents' preferences for proximity) is employed. However, ALMA is a more complex model that allows exploration of various research questions related to land markets. The model was also applied to study the effects of both environmental amenities and disamenities on land prices and land patterns in a heterogeneous landscape, the effects of heterogeneous flood risk perception on land values and development in high-risk zones in a coastal land market, the land market response of economic agents to climate change, and parameterization of economic agents with micro-level data from a survey conducted in the Netherlands [21], and the effects of a land tax designed to preserve ecosystems on land patterns [22]. The general conclusions from our different modeling exercises are that agent-based land market models run under the assumption of homogeneous agents and homogeneous landscapes replicate the results of conventional analytical models in urban and environmental economics. However, if assumptions of agent or spatial

[9] A formal statistical test of the difference in significance between these estimated coefficients between the multiple model runs for both experiments is conceptually possible, but is beyond the scope of this paper.

homogeneity are relaxed, the results become qualitatively different in ways that have substantial policy implications.

In our ABM market, there is no one unique equilibrium-determined price for everyone in the market; rather, there is a set of individual transaction prices determined via bilateral trading by each set of trading partners separately. In spite of the fact that the centralized price determination mechanism is replaced by the distributed bilateral trading, the ALMA model with homogeneous agents reproduces the qualitative results of a monocentric urban model.

In the case of heterogeneous individual preferences for proximity, the land price gradients no longer exactly follow the predictions of the analytical model. In particular, the land price still generally decreases with distance to the city center, but the prices of the equidistant cells are no longer equal, since individuals with heterogeneous tolerance for commuting value them differently.

The most interesting result is that the city border has expanded due solely to the introduction of heterogeneity in agents' preferences for proximity. Essentially, the existence of agents who are more tolerant for commuting creates a ground for urban sprawl. Thus, heterogeneity among individual location preferences is likely to be one of the factors causing urban sprawl and needs to be accounted in policy development. To our knowledge this result has not been reported before. Empirical econometric modeling [23, 24] has demonstrated the relationship between urban sprawl and landscape heterogeneity (green amenities). Agent-based urban models demonstrate that heterogeneous agents and heterogeneous landscape in combination exacerbate urban sprawl [14]. However, the fact that heterogeneity in preferences per se causes city expansion and spatially heterogeneous land rent patterns is a new result that could be demonstrated only through the agent-based land market, since the standard urban economic models cannot be solved with heterogeneous preferences for a 2D landscape.

The introduction of preference for open-space amenities and/or aversion to urban density has been shown to produce discontinuous patterns of development [14, 25–27]. We expect similar results with the ALMA model when open-space amenities are introduced. We also expect that the combination of heterogeneity of preferences for proximity and a heterogeneous landscape will exacerbate urban expansion and sprawl, especially if the distribution of open-space amenities is modeled in a realistic way. Usually, the level of environmental amenities increases with distance from the CBD. So, those people who are already tolerant for commuting receive additional benefits of settling farther from the city center. These households are willing to pay more for a remote location if it has a scenic view or a park close by, so more open space is converted into urban use and the city expands further. We leave these experiments for future work.

With the help of a simple regression analysis of the model-generated data, we demonstrated that inclusion of data on individual preferences (available in the case of ABM) increases the explained variation in land prices. Essentially, we have created a computational laboratory in which we have a full understanding of the agent-level and spatial factors that influence bid prices, ask prices, and realized transactions. This laboratory lets us explore the statistical predictions that emerge from these models, creating an opportunity for greater understanding of the potential processes that have generated the transaction data that we observe in the real world.

Acknowledgements Funding from NWO-ALW (LOICZ-NL) project 014.27.012 and the US National Science Foundation grants 0414060 and 0813799 is gratefully acknowledged.

References

1. Alonso W (1964) Location and land use. Harvard University Press, Cambridge, MA
2. Strazsheim M (1987) The theory of urban residential location. In: Mills ES (ed) Handbook of regional and urban economics. Elsevier Science Publishers B.V., Amsterdam, pp 717–757
3. Filatova T, Parker D, van der Veen A (2009) Agent-based urban land markets: agent's pricing behavior, land prices and urban land use change. J Artif Soc Soc Simulat 12(1):3. Available online: http://jasss.soc.surrey.ac.uk/12/1/3.html
4. Kirman AP (1992) Whom or what does the representative individual represent? J Econ Perspect 6(2):117–136
5. Axtell R (2000) Why agents? On the varied motivations for agent computing in the social sciences. In: Working paper no 17. Center on Social and Economic Dynamics, The Brookings Institution, Washington, DC
6. Manski CF (2000) Economic analysis of social interactions. J Econ Perspect 14(3):115–136
7. Arthur WB (2006) Out-of-equilibrium economics and agent-based modeling. In: Judd KL, Tesfatsion L (eds) Handbook of computational economics, vol 2. Agent-based computational economics. Elsevier B.V., Amsterdam, pp 1551–1564
8. Tesfatsion L (2006) Agent-based computational economics: a constructive approach to economic theory. In: Judd KL, Tesfatsion L (eds) Handbook of computational economics, vol 2. Agent-based computational economics. Elsevier B.V., Amsterdam, pp 831–880
9. Parker DC, Berger T, Manson SM (eds) (2002) Agent-based models of land-use and land-cover change: report and review of an international workshop, October 4–7, 2001. LUCC report series, vol 6, LUCC Focus 1 office: Bloomington, 140
10. Berger T (2001) Agent-based spatial models applied to agriculture: a simulation tool for technology diffusion, resource use changes, and policy analysis. Agr Econ 25(2–3):245–260
11. Happe K (2004) Agricultural policies and farm structures – agent-based modelling and application to EU-policy reform. IAMO Studies on the agricultural and food sector in Central and Eastern Europe, vol 30
12. Polhill JG, Parker DC, Gotts NM (2005) Introducing land markets to an agent based model of land use change: a design. In: Representing social reality: pre-proceedings of the third conference of the European Social Simulation Association. Verlag Dietmar Fölbach, Koblenz, Germany
13. Filatova T, Parker DC, van der Veen A (2007) Agent-based land markets: heterogeneous agents, land prices and urban land use change. In: Proceedings of the 4th conference of the European Social Simulation Association (ESSA'07), Toulouse, France
14. Brown DG, Robinson DT (2006) Effects of heterogeneity in residential preferences on an agent-based model of urban sprawl. Ecol Soc 11(1):46
15. Grevers W (2007) Land markets and public policy. University of Twente, Enschede, Netherlands
16. Parker DC, Filatova T (2008) A conceptual design for a bilateral agent-based land market with heterogeneous economic agents. Comput Environ Urban Syst 32:454–463
17. Hawksworth J, Swinney P, Gilbert N (2008) Agent-based modelling: a new approach to understanding the housing market. PricewaterhouseCoopers LLP, London

18. Robinson DT, Brown DG (2009) Evaluating the effects of land-use development policies on ex-urban forest cover: an integrated agent-based GIS approach. Int J Geogr Inform Sci 23(9):1211–1232
19. Anas A (1990) Taste heterogeneity and urban spatial structure – the logit model and monocentric theory reconciled. J Urban Econ 28(3):318–335
20. Barreteau O, Bousquet F, Attonaty J-M (2001) Role playing game for opening the black box of multi-agent systems: method and lessons of its application to Senegal River Valley irrigated systems. J Artif Soc Soc Simulat 4(2):12
21. Filatova T (2009) Land markets from the bottom up: micro-macro links in economics and implications for coastal risk management. PhD thesis, University of Twente, Enschede, Netherlands, p 196
22. Filatova T, van der Veen A, Voinov A (2008) An agent-based model for exploring land market mechanisms for coastal zone management. In: Sànchez-Marrè JBM, Comas J, Rizzoli A, Guariso G (eds) Proceedings of the iEMSs fourth biennial meeting: international congress on environmental modelling and software (iEMSs 2008), Barcelona, pp 792–800
23. Irwin E, Bockstael N (2007) The evolution of urban sprawl: evidence of spatial heterogeneity and increasing land fragmentation. Proc Natl Acad Sci USA 104(52):20672–20677
24. Irwin EG, Bockstael NE (2004) Land use externalities, open space preservation, and urban sprawl. Reg Sci Urban Econ 34:705–725
25. Caruso G, Peeters D, Cavailhes J, Rounsevell M (2007) Spatial configurations in a Periurban city. A cellular automata-based microeconomic model. Reg Sci Urban Econ 37(5):542–567
26. Parker DC, Meretsky V (2004) Measuring pattern outcomes in an agent-based model of edge-effect externalities using spatial metrics. Agr Ecosyst Environ 101(2–3):233–250
27. Irwin EG, Bockstael NE (2002) Interacting agents, spatial externalities and the evolution of residential land use patterns. J Econ Geogr 2:31–54

The Agent-Based Double Auction Markets: 15 Years On

Shu-Heng Chen and Chung-Ching Tai

Abstract Novelties discovering as a source of constant change is the essence of economics. However, most economic models do not have the kind of novelties-discovering agents required for constant changes. This silence was broken by Andrews and Prager 15 years ago when they placed GP (genetic programming)-driven agents in the double auction market. The work was, however, neither economically well interpreted nor complete; hence the silence remains in economics. In this article, we revisit their model and systematically conduct a series of simulations to better document the results. Our simulations show that human-written programs, including some reputable ones, are eventually outperformed by GP. The significance of this finding is not that GP is alchemy. Instead, it shows that novelties-discovering agents can be introduced into economic models, and their appearance inevitably presents threats to other agents who then have to react accordingly. Hence, a potentially indefinite cycle of change is triggered.

Keywords Novelties discovering • Economic changes • Double auctions • Genetic programming • Autonomous agents

---

S.-H.C hen (✉)
Department of Economics, AI-ECON Research Center, National Chengchi University, No.64, Sec.2, ZhiNan Rd., Wenshan District, Taipei City 11605, Taiwan (R.O.C)
e-mail:c hen.shuheng@gmail.com

C.-C.T ai
Department of Economics, Tunghai University, No.181, Sec.3, Taichung Harbor Road, Taichung 40704,T aiwan(R .O.C.)
e-mail:c hungching.tai@gmail.com

K. Takadama et al. (eds.), *Simulating Interacting Agents and Social Phenomena: The Second World Congress*, Agent-Based Social Systems 7,
DOI 10.1007/978-4-431-99781-8_9,

1 Introduction: It Takes Time to See "Change"

Economics is about change, and that subject has been clearly stated in Alfred Marshall's following famous quotation:

> Economics, like biology, deals with a matter, of which the inner nature and constitution, as well as outer form, are constantly changing ([18], p. 772).

While "constantly changing" is highlighted frequently in various documents of daily life, it seems that economists have not yet been sure whether they do have a capable model for this subject. In fact, the recent book by Frydman and Goldberg [13] has just affirmed the lack of an adequate economic model for change, which has also been pointed out by Herbert Simon many years ago [22].

For Simon, what matters is the process which leads to constant change and novelties discovering:

> ...if we want to have a theory of technological change, it will have to be a theory of the processes that bring about change rather than a theory of specific nature of the changes [22].

To have those features, the model should be able to constantly generate new opportunities (potential to change), and agents, as part of the model, should be able to constantly exploit these opportunities (potential to novelties discovery). What may or may not come to our surprise is that infinitely smart agents, the *homo economicus*, are not qualified to be constituents of this kind of models. Neither can most adaptive agents used or studied in economics serve this purpose, mainly because most of these adaptive agents are equipped with tools which can only handle *well-structured* problems, not the *ill-structured* ones.[1]

Genetic programming (GP) is one algorithm, although not the only one, which may equip agents with those capabilities.[2] Using the terms of Simon [22], genetic programming is a *chunk*-based search algorithm. These chunks, according to Simon, provide the basis for human agents to recognize patterns and develop intelligent behavior. These chunks may also be known as building blocks [17] or modules [21]. Simon considered that, in addition to a 10-year experience, 50,000 chunks are required to be an expert. These two magic numbers nicely match the two parameters in GP, namely, the *number of evolving generations and the population size*.

Hence, an agent, endowed with a population size of 50,000 "chunks" (chromosomes, building blocks, LISP trees, parse trees), after 10-year equivalent iterations (learning, evolution), can become an expert. This kind of adaptive agent, referred to as the GP-based agents for convenience, provides us with a starting point for

---

[1] See [22], p. 28–30.

[2] Genetic algorithms and learning classifier systems can be other alternatives. However, to the best of our knowledge, most agent-based economic applications of genetic algorithms do not manifest this capability, and, for some reason not exactly known, there are almost no agent-based economic applications of learning classifier systems.

modeling change and novelties discovery. One of the best demonstrations is the use of GP in the agent-based double auction markets.[3]

The rest of this paper is organized as follows. Section 2 provides a literature review. Section 3 presents the experimental design. The simulation results are analyzed and discussed in Sect. 4, followed by the conclusion in Sect. 5.

2 Agent-Based Double Auction Markets: Literature Review

In the double auction market, both sides of the market (buyers and sellers) are able to submit prices, bids from buyers and asks from sellers, to signify how much they want to buy or sell for certain number of units of the trading target. The bids and asks are then matched by first ranking them in descending order and ascending order, respectively. If the highest bid is greater than the lowest ask, then the transaction can happen, and the price can be settled somewhere between the bid and ask, say, in the middle. The matching will continue until all remaining bids are smaller than remaining asks; till then, and a round of matching is over. All unfinished or potential trade can be submitted in the next round with possibly more competitive or attractive bids and asks. Round after round, the market can continue indefinitely.

This double auction mechanism has been practically applied to many markets. The pit of the Chicago commodities market is an example; the New York Stock Exchange, another. This market mechanism also inspired the earliest idea of economic experiments [23], and was shown to be very efficient in achieving the equilibrium price. Such a result, in a sense, nicely confirms the well-known Adam Smith's *invisible hand* or the *Hayek hypothesis* [16].

2.1 Gode-Sunder Model

Since this market was shown to be so efficient, whatever individual traders actually knew, learned or did during the trading process was considered completely irrelevant. Gode and Sunder were thus motivated to test a hypothesis that intelligence is completely irrelevant to the market efficiency of the double auction market by proposing what is known as *zero-intelligence agents* [15]. Not only are these agents unable to learn, they basically behave completely randomly. Gode and Sunder showed that this kind of zero-intelligence software agent could perform as well as human agents in the double auction experiments.

Gode and Sunder [15] is one of the earliest agent-based double auction markets, while back in the early 1990s, the term *"agent-based computational economics"*

[3]The reason why we choose the agent-based double auction market as the main pursuit of this paper is because this is one of the few economic models in which human agents, programmed agents and autonomous agents have been involved. See Sect. 2 for the details.

(ACE) has not yet appeared. Nonetheless, the elements of ACE were in the Gode-Sunder simulation model, mainly from the specification of the behavioral rules of software agents to the emergent outcome through the interactions of these agents. In [15], these software agents simply behave randomly; yet the emergent outcome was a highly efficient market. This result was quite surprising.

> Adam Smith's invisible hand may be more powerful than some may have thought; it can generate aggregate rationality not only from individual rationality but also from individual irrationality ([15], p. 119).

Per [15], the invisible hand even exists in a market composed of non-purposive agents (individual irrationality). However, our Homo Sapiens are definitely purposive. When placed in a well-defined experiment like the double auction market, Homo Sapiens are naturally attracted by transaction gains, and it is not likely that blind bidding and asking is a sensible way to react to the information they acquired.[4]

2.2 Santa Fe Double Auction Markets

The purposive traders not only will not bid or ask randomly, but may even develop some strategies to trade, be they sophisticated or simple. In fact, an inquiry into the effective characterization of the "optimal" trading strategies used in the double auction market led to a series of tournaments, known as the *Santa Fe Double Auction Tournament* [19, 20].[5] This tournament organized by the Santa Fe Institute invited participants to submit trading strategies (programs) and tested their performance in comparison with other submitted programs in the *Santa Fe Token Exchange*, an artificial market operated by the double auction mechanism. More than 20 programs based on different design principles were proposed, and the best-performing one was the Kaplan program.[6]

The Santa Fe Double Auction (SFDA) Tournament provides another early example of the agent-based double auction markets. Differing from the Gode-Sunder model, SFDA considers software agents strategic but also hand-written by Homo Sapiens. This design gives the software agents a dual role. On the one hand, they are programmed agents (machine codes); on the other hand, they are incarnations of Homo Sapiens. The subtle difference between the two lies in the decision made *on-line* vs. *off-line*. Human agents make on-line decisions. They receive immediate feedback, but are pressed to react. Human-written programs are generated off-line, so time pressure is not imminent; however, participants receive no immediate feedback while writing their programs. Therefore, the program writing

[4]As we shall see below, zero-intelligence agents or slightly modified zero-intelligence agents cannot compete with some well-thought human-written programs.

[5]The first DA tournaments were held by the Santa Fe Institute in 1990. A share of $10,000 was offered to the writers of algorithms that could perform well in a double auction competition. The tournament attracted around 25 different and well thought-out strategies.

[6]Submitted by Todd Kaplan, then a student at the University of Minnesota. See Sect. 3.2.

relies largely on the participants' mind power and is more like a *deductive process*. Accordingly, double auction experiments and double auction tournaments provide us with two different ways to observe the human decision-making process. The on-line decision is more inductive, and, possibly, simple but spontaneous, while the off-line decision is more deductive, and, possibly, complex but less adaptive.

2.3 Andrews-Prager Model

Andrews and Prager [1] integrated both the human agents in experimental markets and the software agents in agent-based double auction markets. In [1], software agents were randomly generated by using the *initial knowledge* (the *primitives*, the *building blocks*) inspired by the human-written program.[7]

What Andrews and Prager did was to make the computer first randomly generate trading programs; in this sense, it was similar to Gode and Sunder's zero-intelligence agents. However, only in the very beginning were these programs truly randomly generated. After that, these programs were placed in agent-based double auction markets with other software agents, e.g., software agents from SFDA, and then tested, reviewed and revised based on their performance. Some new programs would be generated after that. Nevertheless, this further generation was no longer random, but biased toward the revision of the existing well-performing programs, and the deletion of the ill-performing ones. This brought the on-line learning to the software agents (or programmed agents) and made them become *autonomous agents* so that they could behave like human agents in the experimental markets, in terms of spontaneous and fast reacting.

By using genetic programming to generate these autonomous agents, Andrews and Prager [1] were the first to apply genetic programming to double auction markets. Their model is briefly sketched in Fig. 1. What Andrews and Prager did was to fix a trader (Seller 1 in their case) and used genetic programming to evolve the trading strategies of only that trader. In the meantime, one opponent was assigned the trading strategy "Skeleton", a strategy prepared by the SFDA tournament. The trading strategies of the other six opponents were randomly chosen from a selection of the submissions to SFDA. Such a design was to see whether GP could help an individual trader to evolve very competitive strategies given their opponents' strategies.

The simulation model established by Andrews and Prager [1] enables us to move one step toward a genuine economic model of change. The key ingredient is the autonomous agent, driven by genetic programming. These autonomous agents, by design, are purported to search for better deals to gain from. In the very foundation of classical economics, these agents (autonomous agents) contribute to the discovery

[7]More details will be given in Sect. 3.3. In brief, all randomly generated programs can be regarded as samples from the *span* of some *bases*. These bases, as listed in Table 1, are all from human-written programs.

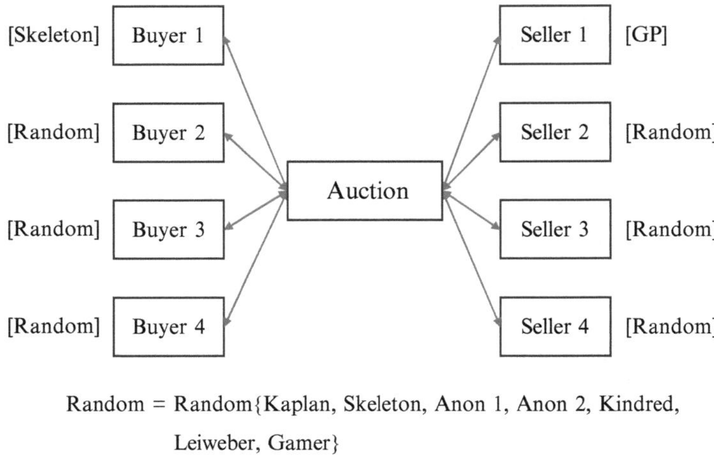

Fig. 1 The Andrews-Prager double auction model

and exploitation of hidden patterns and opportunities. Their reactions further lead to the change of economy, which in turn create new opportunities for further exploitation. This indefinite cycle is an essential, if not the whole, part of Alfred Marshall's biological description of economy as an "*constantly changing*" [18].

Despite this great potential to interest economists, Andrews and Prager's interpretation of their model was rather less telling, and failed to draw the attention of those economists who have little background in social simulation. Besides, their agent-based model was neither fully constructed nor extensively simulated. Only one market participant, instead of all, is autonomous. This certainly restricts the extent of endogenous change, which a genuine model of economic change may have. Other than that, only few experiments have been attempted, and their statistics were not well presented. There was no further development of this model. Hence, the work on the agent-based double auction market, as a genuine model of change, ceased until the late 1990s, when this model was finally revisited by two economists, Herbert Dawid [10] and Shu-Heng Chen [3].

In the following, we shall only review the work by Chen and his colleagues, because the series of Chen's work can be regarded as a direct extension of the Andrews-Prager model. We shall call this later-developed agent-based double auction market AIE-DA (standing for AI-ECON double auction) to distinguish it from SFDA and the Andrews-Prager model.

2.4 AIE-DA

The AIE-DA is probably the only agent-based double-auction market which has received extensive and systematic study. Chen and Tai [5] first extended the Andrews-Prager model by making all market participants autonomous. This is also done by applying genetic programming, as shown in Fig. 2. The architecture of the

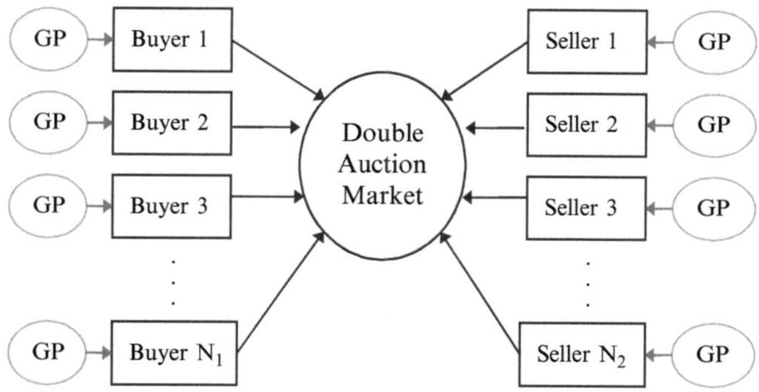

Fig. 2 The AIE-DA double auction market

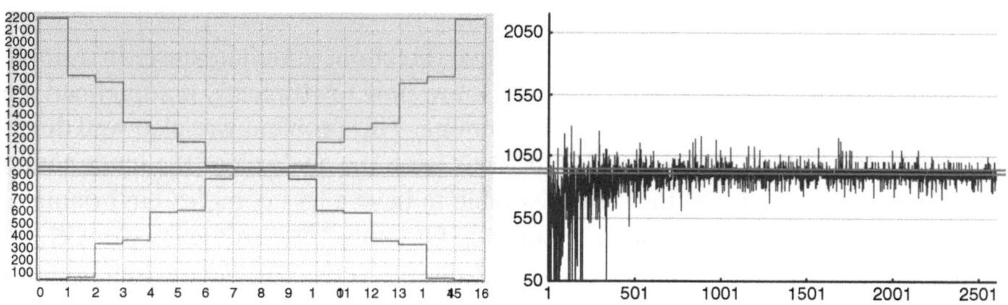

Fig. 3 Agent-based double auction market simulation with MGP agents

genetic programming used is known as *multi-population genetic programming* (MGP). In brief, they viewed or modeled each agent as *a single population of bargaining strategies*. Genetic programming is then applied to evolve each population of bargaining strategies. In this model, a society of bargaining agents consists of many populations of programs.

Chen and Tai [5] first showed that the market composed of this kind of autonomous agent could exhibit behavior similar to what we learn from market experiments with human subjects [23]. Figure 3 demonstrates a typical result observed in this agent-based double auction market. The left panel of the figure is the simulated market environment defined by the demand and supply schedule, whereas the right panel of the figure gives the price dynamics resulting from the bargaining behavior generated by MGP agents. As we can see from Fig. 3, market prices quickly move toward the equilibrium price (or price interval), and then slightly fluctuate around there. The AIE-DA model's successful replication of the market experiment results motivated us to return to the work initiated, while largely unfinished, by Andrews and Prager.

Our research question is: given a set of opponents, regardless of who they are, and how smart they are, as long as they are non-autonomous, can our genetic-programming agent eventually outperform them? This, in our opinion, is the fundamental issue in a discipline where change is her sole concern and no-arbitrage state is the consequence of change. Notice that when opponents are non-autonomous, their behaviors are largely certain, even in a stochastic sense. This in turn implies that, unless they are perfect, there is always a way to outperform them. Since our autonomous agents, by design, are constantly looking for chances, opportunities, and patterns, finding a way to outperform them should be just a matter of time. Andrews and Prager [1] also had this conjecture, but they did not move far enough to document a proof. We, therefore, go back to where Andrews and Prager started, while clothed with the legacy of Alfred Marshall or Charles Darwin, to see whether we can return the missing element, autonomous agents, to economics.

This research question can be further separated into two different directions: using Alfred Marshall's term, inner nature (constitution) and outer form. The former focuses on the novelties discovered by the autonomous agents which can help them stand in an advantageous position, whereas the latter refers to their observable performance. Putting the two together, we inquire what will make them perform well. In fact, by observing and understanding what our autonomous agents learned, we as outsiders are also able to learn.

However, it can be hard to tackle these two directions simultaneously in a single study, mainly because we still do not quite know how to efficiently comprehend the "knowledge" generated by genetic programming. This problem was also well documented in [6] and [7]. Therefore, if our focus is on the analysis of the inner nature of the autonomous agents, then it is desirable to have a less complex environment, in other words, less sophisticated opponents. Of course, it also means we will not be able to fully test our autonomous agents. Alternatively, if we put our autonomous agents in a more complex environment with more sophisticated opponents, then it would become much harder to trace how they beat these opponents if they behave so. Therefore, we generated two series of studies to deal with these two directions. Chen et al. [7, 8] are devoted to the analysis of what our autonomous agents discover when they outperform their opponents, whereas this paper is devoted to the second direction.

3 Experimental Design

Experiments in this paper were conducted using the AIE-DA platform. In this double auction environment, similar to [1]'s model, traders can be assigned different trading strategies from the economic literature. Thus GP agents and other software trading strategies are allowed to coexist in the market, and the combination of them together with the random demand-supply arrangements constitute a variety of market conditions in which we can test our autonomous agents.

3.1 Market Mechanism

Our experimental markets consist of four buyers and four sellers. Each of the traders can be assigned a specific strategy – either a non-autonomous trading strategy or an autonomous GP agent. During the trading processes, traders' identities are fixed so that they cannot switch between buyers and sellers.

In this study, we endow our traders by adopting the same token generation process as in [20]'s design. Each trader has four units of commodities to buy or to sell, and can submit only once for one unit of commodity at each step of a trading day.

Since the AIE-DA is a discrete double auction market, it will not clear before receiving every trader's order at each trading step. The AIE-DA adopts the AURORA trading rules such that at most one pair of traders is allowed to make a transaction at each trading step. The transaction price is set to be the average of the winning buyer's bid and the winning seller's ask.

Every simulation lasts 7,000 trading days, and each trading day consists of 25 trading steps. At the beginning of each simulation, traders' tokens (reservation prices) are randomly generated with the random seed 6,453.[8] Therefore, each simulation starts with a new combination of traders and a new demand-supply schedule. At the beginning of each trading day in a specific simulation, every trader's tokens are replenished. Thus the AIE-DA is in fact a repeated double auction trading game.

3.2 Trading Strategies

In order to extensively test whether our autonomous trading agent can exhibit its learning capability, various kinds of trading strategies were collected from the double auction literature and were injected into the markets as GP agents' competitors[9]:

- *Truth Teller*: Truth-telling traders simply bid/ask with their reservation prices.
- *Skeleton*: The Skeleton strategy was the strategy provided to all entrants of the Santa Fe Double Auction (SFDA) Tournament as a reference material [20]. The Skeleton strategy simply bids or asks by referring to its own reservation prices and the current bid or the current ask in the market.
- *Kaplan*: The Kaplan strategy was designed and submitted to the SFDA Tournament by economist Todd Kaplan [20]. It is a so-called "background trader" strategy in the sense that it remains silent until the market bid and the market ask are close enough to imply a trading opportunity. When this opportunity emerges, the Kaplan trader will jump out and steal it. In spite of the simplicity of its tactic, the Kaplan strategy turned out to be the winner of the SFDA Tournament.
- *Ringuette*: Submitted to the SFDA Tournament as well, the Ringuette strategy was designed by computer scientist Marc Ringuette. It is also a background

[8]Please refer to [20] for the random token-generation process.

[9]Named by or after their original designers, these strategies were modified to accommodate our discrete double auction mechanism in various ways. They were modified according to their original design concepts as much as possible. As a result, they might not be 100% the same as their original forms.

trader, whose strategy is to wait until the first time when the current bid exceeds the current ask less a profit margin. The Ringuette strategy is a simple rule of thumb, and it won the second place in the SFDA tournament [20].

- *ZIC (Zero-Intelligence Constrained)*: The ZIC traders were proposed by Gode and Sunder [15]. ZIC traders send random bids or asks to the market in a range bounded by their reservation prices. Although ZIC traders can avoid transactions which incur losses, they don't have any goals or tactics during the trading process. Therefore, they are regarded as "zero-intelligence".
- *ZIP (Zero-Intelligence Plus)*: The ZIP strategy is derived from [9]. A ZIP trader forms bids or asks with a chosen profit margin, and it will try to raise or lower its profit margin by inspecting its own status, the last shout price, and whether the shout prices are accepted or not. Once the profit margin is chosen, the ZIP trader will gradually adjust its current shout price to the target price.
- *Markup*: The Markup trading strategy is drawn from [24]. Markup traders set up certain markup rates and consequently determine their shout prices. In this paper, the markup rate was set to be 0.1.[10]
- *GD (Gjerstad-Dickhaut)*: The GD strategy is proposed by Gjerstad and Dickhaut [14]. A GD trader scrutinizes the market history and calculates the possibility of successfully making a transaction with a specific shout price by counting the frequencies of past events. After that, the trader simply chooses a price as her bid/ask if it maximizes her expected profits.
- *BGAN (Bayesian Game Against Nature)*: The BGAN strategy was proposed by Friedman [12]. BGAN traders treat the double auction environment as a game against nature. They form beliefs in other traders' bid/ask distributions and then compute the expected profit based on their own reservation prices. Hence their bids/asks simply equal their reservation prices minus/plus the expected profit. Bayesian updating procedures are employed to update BGAN traders' prior beliefs.
- *EL (Easley-Ledyard)*: The EL strategy was devised by Easley and Ledyard [11]. EL traders balance the profit and the probability of successfully making transactions by placing aggressive bids or asks in the beginning, and then gradually decrease their profit margin when they observe that they might lose chances based on other traders' bidding and asking behavior.
- *Empirical*: The Empirical strategy was inspired by Chan et al. [2]'s empirical Bayesian traders. The Empirical trader works in the same way as Friedman's BGAN but develops its belief by constructing histograms from opponents' past shout prices.

These strategies are chosen because they can represent, to a certain degree, various types of trading strategies observed in financial market studies. Some of them are simple rules of thumb, such as the Kaplan, ZIP, or EL strategies, while the others are quite sophisticated in their decision processes, such as the GD, BGAN, and

---

[10]We choose 0.1 because [24]'s simulations shows that the market efficiency will be maximized when traders all have 0.1 markup rates.

Empirical strategies. From the viewpoint of adaptivity, some of them are adaptive in the sense that they adjust in response to the market situations, while the others are non-adaptive by repeating the same behavior regardless of the environment.

Despite their distinct features, none of these strategies is autonomous because their trading tactics are predefined according to some fixed principles. In the following section, we will introduce our autonomous trading agent, whose principle is to constantly exploit the environment and to look for the fittest behavior at the time.

3.3 GP Trading Agents

As introduced in Sect. 2.4, each GP trader in AIE-DA comprises a number of bargaining strategies which can be represented by parse trees. We provide GP traders with basic market, as well as private, information and a set of elementary operators and functions, so that they can construct their strategies.[11] Table 1 explains all such primitives available to the GP trader.

Table 1 The terminal and the function sets of the GP traders

Terminal set	
PMax, PMin, PAvg	The maximum, minimum, and average prices for the previous day
PMaxBid, PMinBid, PAvgBid	The maximum, minimum, and average bids for the previous day
PMaxAsk, PMinAsk, PAvgAsk	The maximum, minimum, and average asks for the previous day
CASK, CBID	The highest bid and the lowest ask in the previous trading step
HT, NT, LT	The first, next, and last reservation prices owned by each trader
TimeLeft, TimeNonTrade	The number of steps left in this trading day, and the number of consecutive no-transaction steps until the current step
Pass, Constant	To give up bidding/asking in this step, or to shout a random number
Function set	
+, −, *, /	Basic arithmetic operations to add, subtract, multiply, or divide
Abs, Log, Exp, Sin, Cos, Max, Min	Basic mathematical functions
If-Than-Else, If-Bigger-Than-Else, Bigger	Basic logical operators

[11] The elements in the terminal and function sets are extracted from the Skeleton, Kaplan, and Ringuette strategies, which are human-designed trading rules and are proved to be quite efficient in gaining profits. Please refer to [19, 20] for the structure and the performance of these strategies.

We do not train our GP traders before sending them to the double auction tournament. Instead, we provide the GP traders with randomly generated strategies at the beginning of each experiment. This implies that our autonomous GP agents do not have any prior knowledge or experiences to refer to. All they can do is to test and to explore as many of the possibilities as they can on their own.[12]

At the beginning of every trading day, each GP trader randomly picks a strategy from its population of strategies and uses it throughout the whole day. The performance of each selected strategy is recorded. If a specific strategy is selected more than once, its weighted average will be recorded.[13]

GP traders' strategies are updated – with selection, crossover, and mutation – every *N*days, where *N* is called the "select number".[14] Only standard crossover and mutation are performed when the GP trader renovates its strategies, which means that no election, ADFs (Automatically Defined Functions), or other mechanisms are implemented. When choosing the parents for the next-generation strategies, the tournament selection is implemented and the size of the tournament is 5, regardless of the size of the population. We also preserve the elite for the next generation, and the size of the elite is 1. The mutation rate is set at 5%, of which 90% is tree mutation.[15]

3.4 Experimental Procedures

Since we have only eight traders (four buyers and four sellers) in the market while there are 12 trading strategies to be tested, we compare the strategies by randomly sampling (without replacement) those eight strategies and injecting them into the market one at a time. We did not try out all the possible combinations and permutations of strategies; instead, 300 random match-ups were created for each series of experiment. In each of these match-ups, any selected strategy will face strategies completely different from its own kind. For example, a certain type of strategy such as ZIC will never meet another ZIC trader in the same simulation. Thus, there is at most one GP trader in each simulated market, and this GP trader adjusts its bidding/

[12]For a more detailed explanation about how GP can be used to construct trading strategies in double auction markets, i.e., how strategies are generated and renovated with crossover and mutation, please refer to [4].

[13]The fitness value of GP traders is defined as the achievement of the individual efficiency, which will be explained later in Sect. 4.

[14]To avoid the flaw that a strategy is deserted simply because it is not selected, we set *N* as twice the size of the population, so that theoretically each strategy has the chance to be selected twice.

[15]The tournament size and the mutation rate are two important parameters which may influence GP traders' performance. On the one hand, the larger the tournament size, the earlier that the convergence of strategies can be expected. On the other hand, the larger the mutation rate, the more diverse the genotypes of the strategies are. When facing a dynamic problem such as making bids/asks in a double auction market, the impact of different tournament sizes together with different mutation rates on GP performance can only be accessed with a comprehensive experimentation of different combinations. Generally speaking, in many studies the size of the tournament ranges from 2 to 5, while the mutation rate ranges from 1 to 10%.

asking behavior by learning from other kinds of strategies. There is no co-evolution among GP traders in our experiments.

In order to extensively test the capability of our autonomous GP agents, the ten multi-agent experiments were conducted by setting the GP traders' population sizes to be 5, 20, 30, 40, 50, 60, 70, 80, 90, and 100, respectively. In each simulation, the same market demand and supply are chosen and kept constant throughout the 7,000 trading days.[16] In each trading day, buyers' and sellers' tokens are replenished so that they can start over for another 25 trading steps.

4 Results: GP Agents Versus Non-Autonomous Traders

In order to evaluate each trader's performance in terms of profits, we adopt the notion of *individual efficiency*. Considering the inequality of each agent's endowment due to the randomized match of strategies as well as the randomized reservation prices, direct comparisons of raw profits might be biased since "luck" may play a very significant role. To overcome this problem, a general index which can evaluate traders' *relative performances* in all circumstances is necessary. The idea of individual efficiency meets this requirement.

The individual surplus, which is the sum of the differences between one's intramarginal reservation prices and the theoretical market equilibrium price, measures the potential profit that a trader can make in the market. Individual efficiency is then calculated as the ratio of one's actual profits to its individual surplus, and thus measures the ability of a trader to explore its potential interests in various circumstances.

In this section, we also evaluate traders' performances from the profit-variation perspective. In addition to profits, a strategy's profit stability is also taken into account because, in double auction markets, the variation in profits might be considered in human trading strategies, which are determined by the human's risk attitudes. In this paper, we procure variations in strategies by calculating the standard deviation of each strategy's individual efficiencies.

To investigate whether our autonomous GP traders are capable of outperforming non-autonomous ones, we first sample GP traders with population size of 5, 50, and 100 denoted as P5, P50, and P100, respectively, and illustrate the results in Fig. 4.

By observing the GP traders' performances, we can clearly answer the following two questions: (1) Can GP traders defeat other strategies? (2) How many resources are required for the GP trader to outperform other strategies?

[16]As mentioned in Sect. 1, we can use GP to model the learning process of an expert possessing intelligence based on, say, a 10-year experience and 50,000 "chunks". In this article, each GP trader has to develop strategies to be used in the markets. These strategies consist of building blocks comprising market variables, and therefore can be viewed as combinations of "chunks". Since we cannot predict how many chunks our GP traders will use, we did not parameterize this variable. Instead, the size of the population of strategies is utilized to characterize this capacity. In a similar vein, we did not model the "10-year experience" directly. 7,000 trading days are available for our GP traders to make their strategies as good as possible.

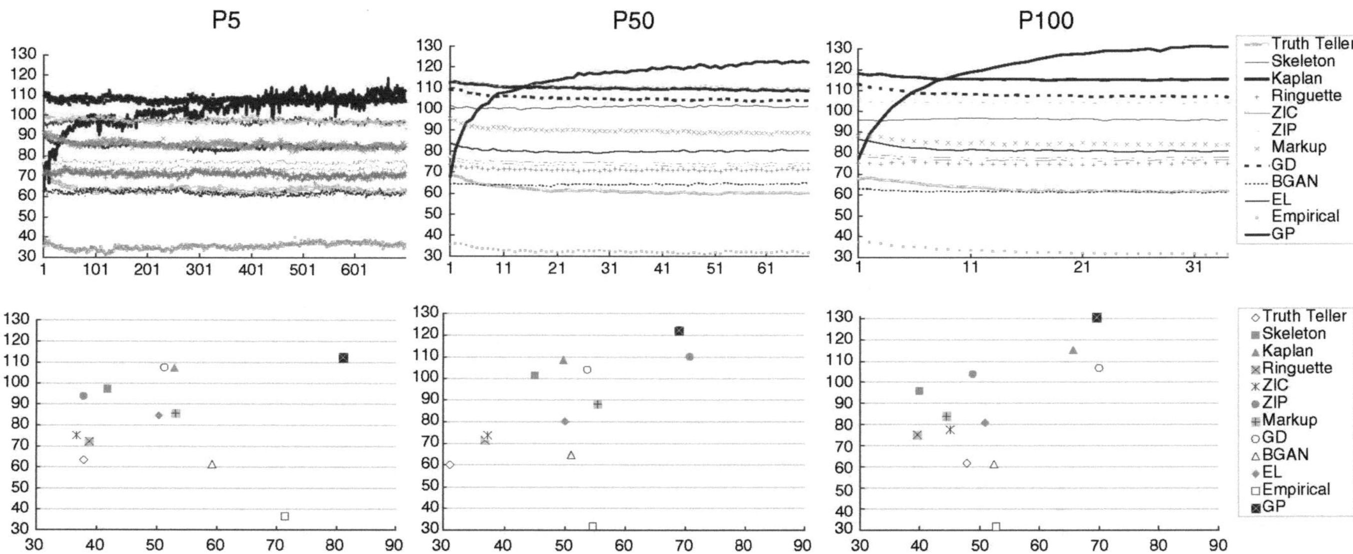

Fig. 4 Comparisons of GP traders with non-autonomous strategies. (**a**) The *top row* are the time series of the individual efficiencies of all the traders. (**b**) The *bottom row* are their profit-variation evaluations in the final generation. The *vertical axis* denotes the individual efficiency, in percentage terms; the *horizontal axis* denotes the standard deviation of their individual efficiency

First, although some of the non-autonomous trading strategies are adaptive in the sense that they can adjust themselves according to the market situations, none of them exhibits an upward trend in terms of performance. In contrast with the apparent growing performances of GP agents, the performances of the non-autonomous strategies are relatively flat. On the other hand, GP traders are able to gradually improve and to outperform other strategies, even under the extreme condition of a population of only 5.[17]

Second, Fig. 4 also illustrates the results in terms of a profit-variation framework. Other things being equal, a strategy with higher profit and less variation is preferred. If we draw a frontier connecting the most efficient trading strategies, Fig. 4 shows that GP traders, even although exhibiting more variation in profits, always occupy the ends of the frontiers.

Third, GP agents need a period of time to learn. The bigger the population, the fewer generations needed to defeat other strategies. In any case, it takes GP traders hundreds to more than a thousand trading days to achieve good performances.[18]

Figure 5 presents a more complete sampling from the GP traders with different population sizes and provides evidence that GP traders with population sizes of

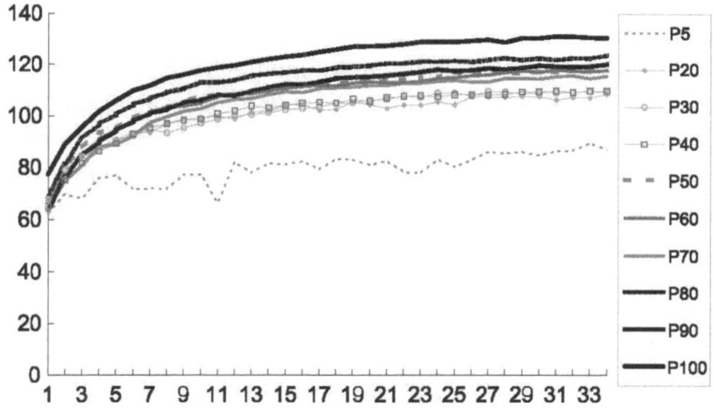

Fig. 5 The improvement in GP performance in generation 34. Generation 34 is chosen because it is the last generation of GP traders with a population size of 100

[17]In our results, the best strategy of GP traders with a population size of 100 in the 34th generation is the selling strategy–*Max(PMinBid, PAvg, PAvgAsk, LT)*, a rather simple rule which adjusts to the market situations by simply choosing whichever is bigger among several types of market and private information. For a more thorough investigation of the kinds of strategies our GP traders are capable of evolving, please see [7].

[18]However, the correlation between the population size and the generations needed to defeat other strategies may not prevail in all circumstances. A GP trader in our double auction tournament is a specific-purpose machine which seeks to discover efficient trading strategies. In such a specific problem where the number of potentially efficient strategies is finite, employing too many strategies (say, 10,000 strategies) may not be coupled with a corresponding increase in the learning speed. In fact, a closer look at our data suggests a decreasing correlation between the population size and the generations needed to defeat the rivals when the population sizes become larger and larger.

5 and 100 constitute the slowest and quickest learners, respectively, while other GP traders lying in between these two enjoy guaranteed performances better than that of P5. These results confirm the superiority of GP traders, and imply that GP traders tend to learn faster when they have larger populations.

5 Conclusion

Novelties discovering as a source of constant change is the essence of economics. In this paper, we propose that a proper model of the constantly changing economies should possess the feature of creating endless opportunities for their participants. This feature, in turn, largely depends on whether the market participants are capable of constantly exploiting such opportunities.

To demonstrate this point, an agent-based double auction tournament inherited from a series of previous studies is launched. This market allows autonomous agents and other non-autonomous trading strategies to pursue their mission of obtaining profits from market transactions. We then extensively test our autonomous agents with different levels of capacities under various kinds of conditions, including different types of opponents and various demand-supply arrangements.

From the results, we can see that our autonomous GP agents are able to dominate the market after a period of learning. The GP traders can outperform prominent opponents such as the Kaplan strategy, which is a simple rule of thumb, and the GD strategy, which has a sophisticated design for the purpose of optimization.

This result suggests that, unless the non-autonomous trading strategies are perfect, there are always chances to take advantage of them. The strategies constructed by our autonomous GP traders, albeit naive in the beginning and simple at the end, are capable of exploiting such opportunities. By continually looking for chances, they gradually become experts in capturing the patterns in their markets.

The significance of this finding shows that novelties-discovering agents can be introduced into economic models, and their appearance inevitably presents threats to other agents who then have to react accordingly. Hence, a potentially indefinite cycle of change is triggered. The existence of autonomous agents, in this sense, becomes the driving force behind an endogenously changing economy.

A natural next step is to compare the behavior of our GP traders to that of human subjects. Human players are supposed to be more adaptive than the programmed agents [20]. We can test this with human experiments, and try to identify whether there is a difference between their abilities in terms of novelties discovering in future research.

Acknowledgements The authors are grateful to the two anonymous referees for their suggestions in regard to the former version of this paper submitted to the WCSS 2008 post-conference proceedings. National Science Council Research Grant No. NSC 95-2415-H-004-002-MY3 and National Chengchi University Top University Program No. 98H432 are also gratefully acknowledged.

References

1. Andrews M, Prager R (1994) Genetic programming for the acquisition of double auction market strategies. In: Kinnear KE Jr (ed) Advances in genetic programming. MIT Press, Cambridge, MA, pp 355–368
2. Chan NT, LeBaron B, Lo AW, Poggio T (1999) Agent-based models of financial markets: a comparison with experimental markets. MIT artificial markets project, Paper No. 124, September 5, 1999. Available via CiteSeer http://citeseer.ist.psu.edu/chan99agentbased.html. Cited 18 June 2008
3. Chen S-H (2000) Toward an agent-based computational modeling of bargaining strategies in double auction markets with genetic programming. In: Leung KS, Chan L-W, Meng H (eds) Intelligent data engineering and automated learning-IDEAL 2000: data mining, financial engineering, and intelligent agents. Lecture notes in computer science 1983. Springer, pp 517–531
4. Chen S-H, Chie B-T, Tai C-C (2001) Evolving bargaining strategies with genetic programming: an overview of AIE-DA Ver. 2, Part 2. In: Verma B, Ohuchi A (eds) Proceedings of fourth international conference on computational intelligence and multimedia applications (ICCIMA 2001). IEEE Computer Society Press, pp 55–60
5. Chen S-H, Tai C-C (2003) Trading restrictions, price dynamics and allocative efficiency in double auction markets: an analysis based on agent-based modeling and simulations. Adv Complex Syst 6(3):283–302
6. Chen S-H, Kuo T-W, Hsu K-M (2008) Genetic programming and financial trading: how much about "what we know"? In: Zopounidis C, Doumpos M, Pardalos P (eds) Handbook of financial engineering, Chapter 8. Springer
7. Chen S-H, Zeng R-J, Yu T (2008) Co-evolving trading strategies to analyze bounded rationality in double auction markets. In: Riolo R, Soule T, Worzel B (eds) Genetic programming theory and practice VI. Springer, New York, pp 195–213
8. Chen S-H, Zeng R-J, Yu T (2009) Micro-behaviors: case studies based on agent-based double auction markets. In: Kambayashi Y (ed) Multi-agent applications with evolutionary computation and biologically inspired technologies: intelligent techniques for ubiquity and optimization. IGI Global
9. Cliff D, Bruten J (1997) Zero is not enough: on the lower limit of agent intelligence for continuous double auction markets. Technical Report no. HPL-97-141, Hewlett-Packard Laboratories. Available via CiteSeer http://citeseer.ist.psu.edu/cliff97zero.html. Cited 18 June 2008
10. Dawid H (1999) On the convergence of genetic learning in a double auction market. J Econ Dynam Contr 23:1544–1567
11. Easley D, Ledyard J (1993) Theories of price formation and exchange in double oral auctions. In: Friedman D, Rust J (eds) The double auction market-institutions, theories, and evidence. Addison Wesley, Redwood City, CA, pp 63–97
12. Friedman D (1991) A simple testable model of double auction markets. J Econ Behav Organ 15:47–70
13. Frydman R, Goldberg M (2007) Imperfect knowledge economics: exchange rates and risk. Princeton University Press, Princeton
14. Gjerstad S, Dickhaut J (1998) Price formation in double auctions. Game Econ Behav 22:1–29
15. Gode D, Sunder S (1993) Allocative efficiency of markets with zero-intelligence traders: market as a partial substitute for individual rationality. J Polit Econ 101:119–137
16. Hayek FA (1945) The use of knowledge in society. Am Econ Rev 54:519–530
17. Holland J (1975) Adaptation in natural and artificial systems. University of Michigan Press, Ann Arbor
18. Marshall A (1924) Principles of economics. MacMillan, New York

19. Rust J, Miller J, Palmer R (1993) Behavior of trading automata in a computerized double auction market. In: Friedman D, Rust J (eds) Double auction markets: theory, institutions, and laboratory evidence. Addison Wesley, Redwood City, CA
20. Rust J, Miller J, Palmer R (1994) Characterizing effective trading strategies: insights from a computerized double auction tournament. J Econ Dynam Contr 18:61–96
21. Simon H (1965) The architecture of complexity. Gen Syst 10:63–76
22. Simon H, Egidi M, Viale R, Marris R (1992) Economics, bounded rationality and the cognitive revolution. Edward Elgar Publishing, Cheltenham
23. Smith V (1991) Bidding and auctioning institutions: experimental results. In: Smith V (ed) Papers in experimental economics. Cambridge University Press, Cambridge, pp 106–127
24. Zhan W, Friedman D (2007) Markups in double auction markets. J Econ Dynam Contr 31:2984–3005

A Doubly Structural Network Model and Analysis on the Emergence of Money*

Masaaki Kunigami, Masato Kobayashi, Satoru Yamadera, Takashi Yamada, and Takao Terano

Abstract This paper presents a new social model on the emergence of money from the barter economy. This Doubly Structural Network (DSN) Model consists of social connections among agents and relationships among commodities recognized by each agent. Our DSN Model is advantageous in describe the micro-macro recognition of exchangeability and the emergence of proto-money that is defined as a commodity with general acceptability. We derive a new approximation dynamics from this DSN model of the emergence of proto-money. Using the structural instability of the dynamics, we show that money can emerge from commodities without distinctive properties. Especially by bifurcation analysis of the dynamics, we find that the social network degree of the society is a definitive factor for non-/single-/double-emergence of proto-money.

Keywords Doubly structural network • Emergence of money • Self-organization • Mean-field dynamics • Bifurcation

1 Introduction

A kind of goods called "money" plays a unique role as a medium for exchange in economy and society, even though this has little practical advantage. This paper presents a new social model on the emergence of the money from the barter economy.

---

*Any views expressed herein are solely those of the author's, and do not represent those of the Government or any representative agencies.

M.K unigami (✉)
Tokyo Institute of Technology/Joint Staff College, Ministry of Defense, Japan
e-mail:m kunigami@gakushikai.jp

M. Kobayashi, S. Yamadera, T. Yamada, and T. Terano
TokyoI nstituteof T echnology, Yokohama,J apan
e-mail: masato.gssm@gmail.com; satoru_yamadera@ybb.ne.jp;
tyamada@trn.dis.titech.ac.jp;t erano@dis.titech.ac.jp

K. Takadama et al. (eds.), *Simulating Interacting Agents and Social Phenomena: The Second World Congress*, Agent-Based Social Systems 7,
DOI 10.1007/978-4-431-99781-8_10, © Springer 2010

This new Doubly Structural Network Model consists of inter-agent social networks and inner-agent recognition networks. The self-organization of these inner-agent recognition networks can describe how general acceptability, an essential nature of a medium of exchange, emerges in the society.

We derive approximate dynamics from the doubly structural network model of the money. This dynamics is improved from our previous work [13] in the following two points. (1) The dynamics shows that even a commodity with no distinctive properties can become a medium of exchange. (2) The dynamics enables bifurcation analysis, which shows that the degree of connection of the social network strongly affects the number of exchange media.

1.1 Problems Regarding the Origin/Emergence of Money

In economics, "money" is usually defined by the following functions [5, 6, 15];

- A medium of exchange
- A unit of value
- Storage of value

Amongst these, many economists maintain that money is essentially a medium of exchange [5–7, 15, 18]. For the emergence of money as a medium of exchange, almost all the agents within the society must recognize that this commodity is exchangeable with almost all others. This nature is called "general acceptability". To focus on the most primitive form of money we define "*proto-money*" as a commodity that has general acceptability. Therefore, in this paper the emergence of money means the emergence of proto-money from a barter economy.

Several theories coexist regarding the origins of money, [15]. The first is the "legal theory of money" that states money has its origins in 'fiats' by Kings or governments. The other is the "theory of commodity money" that states money is spontaneously specialized from exchangeable commodities. In recent years the later has become more popular, but the debate is not necessarily settled. In addition, the "metallic theory" maintains that a commodity becomes money due to its appropriate attributes for exchange, and the "non-metallic theory" maintains something can be money despite its attributes. In this paper, the terms such as "metallic theory of commodity money" or "non-metallic theory of commodity money" are used if necessary.

From the following reasons (a)–(c), this paper analyzes the possibility that "proto-money" can emerge under "non-metallic theory of commodity money".

(a) *Primitive moneys often emerged from societies with no government*

A survey [22] shows primitive moneys was often used in societies without a government or ruler. It is also known that in modern times primitive money emerged from a barter economy in prison camps [23] and in the former Soviet Union [20, 24].

(b) *Primitive moneys has extreme diversity*

In survey [22], the outstanding nature of primitive moneys is its large diversity of forms. It is futile to search for "common applicable properties" in the diversity, and difficult to explain that most of these were not money in other areas/eras.

(c) *Money sometimes collapsed or did not emerge*

The Inca Empire used no money [25]. Bartering arose (in the former Soviet Union [20, 24], and in eighteenth century Pennsylvania [28]). These facts support (a) also, and that the model should also support cases where money does not emerge.

This is not a preconception to reject the "metallic theory" or "legal theory". If the emergence of "proto-money" is difficult under "non-metallic theory", the "metallic theory" is indirectly supported. If the "metallic theory" is also difficult, the "legal theory" is indirectly supported.

1.2 Mathematical Models for the Emergence of Money

The emergence of money is studied in not only economics but also mathematical models. Some research [3, 9, 16] shows that specialization in the exchange media to a certain commodity (e.g. cheapest preserving) is a rational equilibrium strategy in bartering three commodities. An evolutionary model [14] shows that using the cheapest preserving commodity is sometimes unique rational equilibrium also. A matchmaking model of commodities [26] shows that commodity money may spontaneously emerge as the one with the lowest transaction cost. Such research has different approach to ours in Sect. 1.1, since they assume the "metallic theory of commodity money" depending on the particular natures of commodities.

Here are some research examples consistent with the "non-metallic theory of commodity money". A search model for transaction partners [7] shows that bartering and monetary economics are different equilibria, and that monetary equilibrium requires the common recognition that a particular one is money. The evolution of money requires a large fluctuation to break the bartering equilibrium. A simulation model of exchanging commodities [30] shows that by adding the Maxim 'Accept what others accept!' (Menger's 'salability'), a commodity-money emerges when the 'threshold of exchange' in the 'view vector' of the Maxim exceeds a certain level. Another agent simulation [29] in lattice space shows a commodity becomes money based on the "trust" from agents.

In [30] and [26], they point out that the general acceptability of a certain commodity can be represented as a star-shaped network of acceptability or exchangeability around the commodity. We focus on the fact that the networks of exchangeability are essentially important for a bottom up approach. In the next section, we introduce our "Doubly Structural Network Model". This model can describe an emerging process of the general acceptability via coherent growth of individual star-shaped networks of recognition.

From another point of view, the aforementioned research suggests that the emergence of money needs some structural change (change of the 'threshold of exchange' in the 'view vector' [30]), establishing common recognition and a large fluctuation [7], establishing trust [29]) in the society. Following sections illustrate that our model is useful to describe an emergence mechanism based on a social structure.

2 Doubly Structural Network Model

This section introduces a new model of social learning on the social network. This model is unique from other related work since it has double structure of the inter-agent social network and inner-agent recognition networks. This double structure of networks enables us to describe and to analyze the emergence of common knowledge or organized/collective recognitions in the society.

Several models for agents' behaviors and their propagation in a society are known as spatial evolutionary game, infection in network [17, 21, 22], dissemination of culture [1, 12], the Sugar-scape [4], and TPM (Tensor Product Model) [8]. We proposed the Doubly Structural Network Model [13] that handles social propagation of agents' knowledge and recognition such as exchangeability or acceptability of commodities. The structure of this model is illustrated in Fig. 1, and is defined by formula (1).

Similar to the Tag model (Dissemination of Culture) on a network [12], our model is different from TPM [8] that describes the social relationship between agents using not inter-agents connection but "groups". In contrast to the Tag (Dissemination of Culture) model, in our model the propagation of inner representation is not driven by whole similarity of tags but local structural similarity of inner networks. Our model is advantageous in that it can describe not only autonomous structural change but also emergence of structure through simple representation. Here "autonomous structural change" means that each agent's inner network changes depend on not only a neighbors' inner networks but also on its own topology of connection. Also "emergence of structure" means a self-organization of inner networks each of which represents individual recognition on exchangeability.

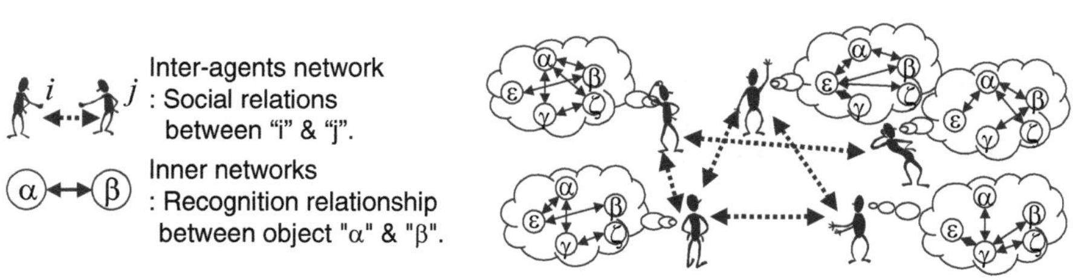

Fig. 1 A doubly structural network of society

$$
\begin{cases}
G^S \equiv \left(V^S, E^S\right), \quad V^S \equiv \left\{v_i^S \middle| i = 1 \sim N\right\}, E^S \subseteq V^S \times V^S \cdots \text{(a)} \\
G_i^I \equiv \left(V^I, E_i^I\right), \quad V^I \equiv \left\{v_\alpha^I \middle| \alpha = 1 \sim M\right\}, E_i^I \subseteq V^I \times V^I \cdots \text{(b)} \\
G^D \equiv \left\{\left\{\left(v_i^S, G_i^I\right) \middle| i = 1 \sim N\right\}, E^S\right\} \qquad \cdots\cdots\cdots \text{(c)} \\
G_{t+dt}^D \equiv F\left(t, G_t^D\right) \qquad \cdots\cdots\cdots \text{(d)}
\end{cases} \tag{1}
$$

- In formula (1a), "social (inter-agent) network" G^S represents the social relationship between agents. The node (vertex) v^S_i represents the i-th agent. The edge set E^S represents connection or disconnection between these agents.
- In formula (1b), "internal (recognition) network" G^I_i represents the internal landscape or recognition of the i-th agent on certain objects (α, β,…). The node (vertex) v^I_α represents the object α. The edge set E^I_i represents connection or disconnection between those objects in the i-th agent's recognition.

  (Whichever directed/undirected graph is available for social or internal network.)

- Formula (1c) shows that "doubly structural network" G^D is created by attaching (/mounting) each internal (recognition) network G^I_i (i = 1,2,…,N) onto the corresponding node *i* (*i*-th agent) of the social network G^S.
- Formula (1d) shows that a propagation/learning model of the doubly structural is defined by the ways of changing states (connection/disconnection) of agents' internal networks via interaction of agents in the social network. In this paper, we use the doubly structural network model of static society in which the social network does not change autonomously. On the other hand, we refer to it as the model of dynamic society if its social network changes autonomously.

The model structure (1a–d) allows us the following advantages;

1. To directly describe the personal recognition of each internal network by shape
2. Tode fine autonomous evolution into the internal networks
3. To describe the micro/macro interaction among agents with these inner evolutions[1]

However, the above formula of the DSN model is a conceptual one without a particular social learning/propagation mechanism, so we need to implement specific inter-agents and inner-agents' interaction for the emergence of money. It enables us to employ an analytical method to derive interesting results on this important classical problem in economics.

[1]The shape of the social network affects the changes of the internal networks (macro→micro). The interactions between the internal networks formed the social attitudes (micro→macro).

3 Doubly Structural Network Model of the Emergence of Money

Here, we implement a specific mechanism to describe the emergence of money in our model [13]. Upon implementation, the social (inter-agents) network reflects the topology of economical/social relationships between agents (indicated as i, j = 1–N). The agents' inner networks represent their own recognition on the exchangeability between commodities (indicated by α,β,γ = 1–M). Each element of the adjacent matrix of the *i*-th agent's inner network is determined as $e_{\alpha\beta}^{(i)} = e_{\beta\alpha}^{(i)} = 1$ if α and β are exchangeable for the *i*-th agent, or as $e_{\alpha\beta}^{(i)} = e_{\beta\alpha}^{(i)} = 0$ if not.

Among possible stages for the emergence of money, this paper focuses on the emergence of proto-money in which a certain commodity achieves general acceptability in the society. In our doubly structural network model, the emergence of proto-money α is represented as a self-organizing process in which a large population of inner-networks becomes similar star-shaped networks with a common hub α (Fig. 2).

The agents in this model interact with each other in the following manners during each time step.

1. *Exchange*: In the social (inter-agent) network, neighboring agents *i* and *j* exchange commodities α and β with probability P_E, if both of them recognize that α and β are exchangeable (i.e. $e^{(i)}_{\alpha\beta} = e^{(j)}_{\alpha\beta} = 1$). All exchanges are assumed to be reciprocal.
2. *Learning*: The learning process of agents consists of the following four methods.
   - *Imitation*: If an agent *i*'s (let $e^{(i)}_{\alpha\beta} = 0$) neighbor *j* and *j*'s neighbor *j'* succeeded in exchanging α-β, then *i* imitate *j* (i.e. $e^{(i)}_{\alpha\beta} \to 1$) with the probability P_I.
   - *Trimming*: If an agent *i* has cycle recognition of exchangeability (e.g. $e^{(i)}_{\alpha\beta} = e^{(i)}_{\beta\gamma} = e^{(i)}_{\gamma\alpha} = 1$), then the agent *i* will trim its inner-network by cutting randomly one of these cyclic edges with the probability P_T. Such a tendency to avoid cyclic exchanges (Zirkulartausch [19]) is consistent with [9] also.

We consider that these two learning processes are essential for the emergence of money. In addition, we introduce two more subsidiary processes as natural fluctuations.

- *Conceiving*: Even if an agent *i* has no recognition of α-β exchangeability ($e^{(i)}_{\alpha\beta} = 0$), it will happen to conceive that ($e^{(i)}_{\alpha\beta} \to 1$) with the probability P_C.
- *Forgetting*: Vice versa, even if an agent *i* has recognition of α-β exchangeability ($e^{(i)}_{\alpha\beta} = 1$), it will happen to forget this ($e^{(i)}_{\alpha\beta} \to 0$) with the probability P_F.

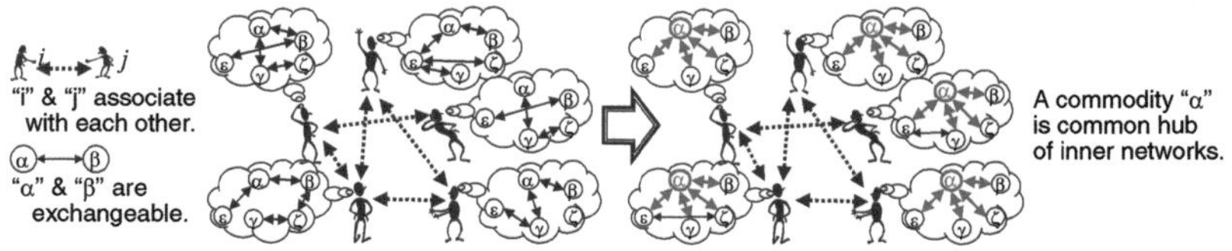

Fig. 2 Emergence of proto-money: common hub among agents represents general acceptability

Although these probabilities are constant data in the model, their values can be dependent on the kind of goods (i.e. $P_E^{(\alpha,\beta)i}$ is not always equal to $P_E^{(\alpha,\gamma)i}$). To simplify the notation, we sometimes omit superscripts$^{(\alpha,\beta)}$ or subscripts$_{\alpha,\beta}$.

4 Mean-Field Dynamics Analysis of the Emergence of Money

4.1 Mean-Field Dynamics

Here, we derive some new mean-field dynamics and analyze the behavior of the doubly structural network model of the emergence of money by mean-field approximation. Mean-field approximation substitutes the overall agent average state for the state around each agent. Instead of ignoring the specific local structure of the network, it makes some analytical approaches possible.

At first, we denote parameter "*k*" as the degree of nodes (agents) on the social network. This *k* is assumed to have certain distribution function *p*(*k*). Next, we introduce the state variable $x_{\alpha;k}$ which represents the average acceptability of commodity α with respect to agents with degree *k*. Each $x_{\alpha;k}$ means the probability that an agent with degree *k* recognizes exchangeability between α and another arbitrary commodity.

The following dynamics describes the time-evolution of these mean-field states.

$$\frac{dx_{\alpha,k}}{dt} = P_E P_I (1 - x_{\alpha,k})k\left(\frac{\sum_{k'} k'(k'-1)p(k')x_{\alpha,k'}}{\sum_{k'''} k'''p(k''')}\left(\frac{\sum_{k''} k''p(k'')x_{\alpha,k''}}{\sum_{k'''} k'''p(k''')}\right)\right)$$
$$- P_T M x_{\alpha,k}^2 \sum_{\beta\neq\alpha} x_{\beta,k} + P_C(1 - x_{\alpha,k}) - P_F x_{\alpha,k}$$
$$0 \le x_{\alpha,k} \le 1 (\alpha = 1,2,\cdots M, k = 1,2,\cdots,N-1). \qquad (2)$$

This consists of four-interaction processes (at 3.2) as shown below.

- First term ~ imitate: A transaction occurs between neighbors and neighbors of neighbors. Here the insides of the medium parentheses represent the expected degree of the neighbors [21] and neighbors' neighbors.
- Second term ~ trim: A cyclic recognition from α to α via β of M commodity types.
- Third terms ~ Conceive and fourth terms ~Forget: Obvious.

4.2 Emergence Scenario Using Mean-Field Dynamics

This section discusses the effect of social network structure on the emergence of proto-money by contriving ideal settings that simplified mean-field dynamics (2).

$$\frac{dx_{\alpha,k}}{dt} = P_E P_I (1-x_{\alpha,k}) k(k-1) x_{\alpha,k}^2 - P_T M x_{\alpha,k}^2 \sum_{\gamma \neq \alpha} x_{\gamma,k} + P_C (1-x_{\alpha,k}) - P_F x_{\alpha,k}$$

$P_{EI}=$ 0.25
$P_T=$ 0.5
$P_C=$ 0.001
$P_F=$ 0.15
$p(k)$ = 1(k=8), 0 (O/W)
$M=$ 32
(α = 1,2,~,32)
dt=0.025

$x_1(0)$= 0.050101
$x_2(0)$= 0.0501
$x_3(0)$= 0.05009
$x_4(0)$~$x_{32}(0)$=0.05

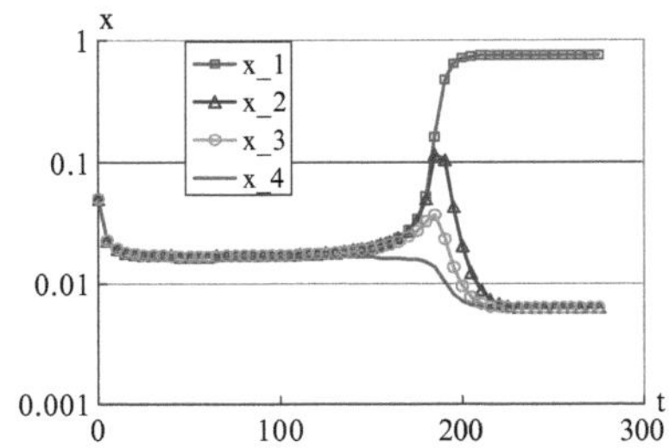

Fig. 3 A numerical outcome of mean-field dynamics in a regular network society shows that proto-money emerges from a homogeneous set of commodities

At first we consider that distribution of the social network has only k-degree node (k has a point distribution ($p(k)=\delta_{k,k0}$), e.g. Regular Network [27]). Although generality is lost, it seems to be appropriately the ideal type for the emergence of proto-money. Figure 3 is one of the numerical outcomes in a society with homogeneous commodities (P*s do not depend on commodities).

This result shows that even though commodity properties are homogeneous, proto-money may emerge spontaneously from infinitesimal fluctuations in the initial condition. This supports the "non-metallic theory of money" that we discussed in Sect. 1.1.

Next, we derive more simplified dynamics to approve the above outcome more generally and to illustrate the effect of social network structure on the emergence of proto-money The "mean-field first-second dynamics" (3) is given by focusing on only the first and the second acceptable commodities and fixing the amount of others in a small constant σ.

$$\begin{pmatrix} \dfrac{dx_\alpha}{dt} = P_E P_I k(k-1)(1-x_\alpha)x_\alpha^2 - P_T M x_\alpha^2 (x_\beta + \sigma) + P_C(1-x_\alpha) - P_F \equiv x_\alpha^2 f_\alpha^{(k)}(x_\alpha, x_\beta) \\ \dfrac{dx_\beta}{dt} = P_E P_I k(k-1)(1-x_\beta)x_\beta^2 - P_T M (x_\alpha + \sigma) x_\beta^2 + P_C(1-x_\beta) - P \equiv x_\beta^2 f_\beta^{(k)}(x_\beta, x_\beta) \end{pmatrix},$$

$$\begin{pmatrix} x_\alpha \equiv x_{\alpha,k}, x_\beta \equiv x_{\beta,k} \\ \sigma \equiv \sum_{\gamma \neq \alpha,\beta} x_{\gamma,K} \ll 1 \\ \alpha,\beta,\gamma = 1,2\ldots M \end{pmatrix}. \tag{3}$$

Assuming enough small σ (total of below the third), the position and stability of the equilibria of the first-second mean-field dynamics (3) shows the non-emergence/single-emergence/double-emergence of proto-money. Global analysis with the isocline method is available for two-dimensional dynamics (3). The isoclines $x_\beta(x_\alpha)$ and $x_\alpha(x_\beta)$ are derived by solving $f^{(k)}_\alpha(x_\alpha, x_\beta)=0$ and $f^{(k)}_\beta(x_\alpha, x_\beta)=0$ in (3).

The isoclines have several shapes according to coefficients (simplifying as $\sigma=0$). At first we describe the bifurcation in the case of "monotonic isoclines".

In Fig. 4, while the degree of the social network k (average number of each agent's neighbors) is small, the equilibrium point Q_1 stays around small level of

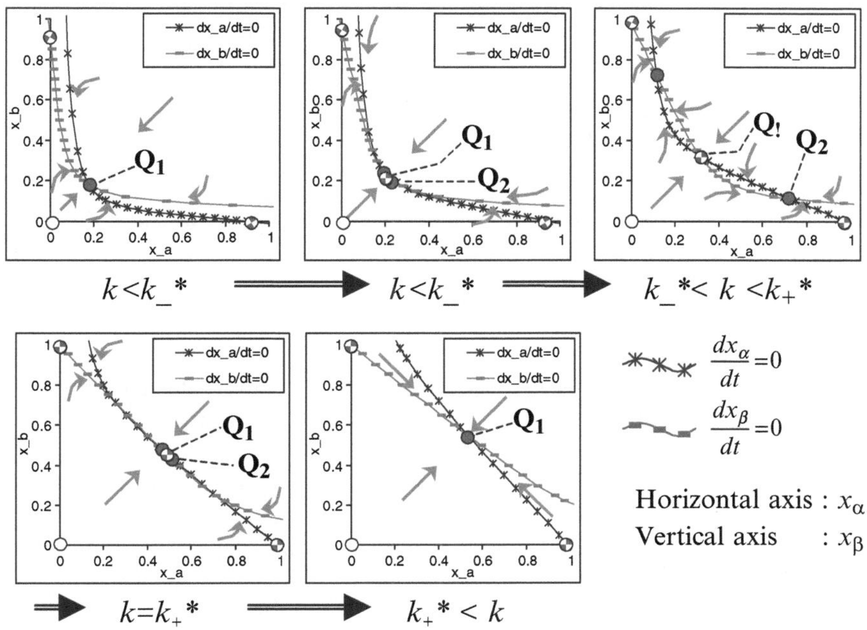

Fig. 4 Bifurcations of first-second dynamics (3) on the case of "monotonic isoclines", the horizontal axis is x_α (largest acceptability of commodity), vertical axis is x_β (second largest one)

acceptability. Once k exceeds a certain value k_-^*(these critical values $k_\pm^*$ are found from the points of tangency of the isoclines), equilibrium splits up, so Q_2 moves towards the area where α takes almost all of acceptability (emergence). Furthermore, if k exceeds k_+^*, the two equilibria are merged, so both of α & β take large acceptability (double emergence).

Similarly, in the case of "isoclines with minima & maxima" and "unimodal (concave) isoclines" (Fig. 5) they have common proto-money emergence scenario as follows:

- *Non-emergence*: no commodities with general acceptability emerge if the degree k is small enough
- *Single-emergence*: only one of the commodities emerges as proto-money if the degree k grows larger than the lower critical value
- *Double-emergences*: Two (or more) commodities emerge as proto-money if the degree k grows larger than the higher critical value

In this way, this doubly structural network model gives us insight into how the social structure affects the emergence of money. This insight explains that proto-money emerges or not as (non-metallic theory of) commodity money based on the structure of a society. Especially this change in the social structure k makes it possible to explain the trigger factors we remarked at in Sect. 1.2. Although the bifurcation parameter k is exogenous in this model, we are conscious of extendibility in which k will increase with rising living standards and decrease with social downfall.

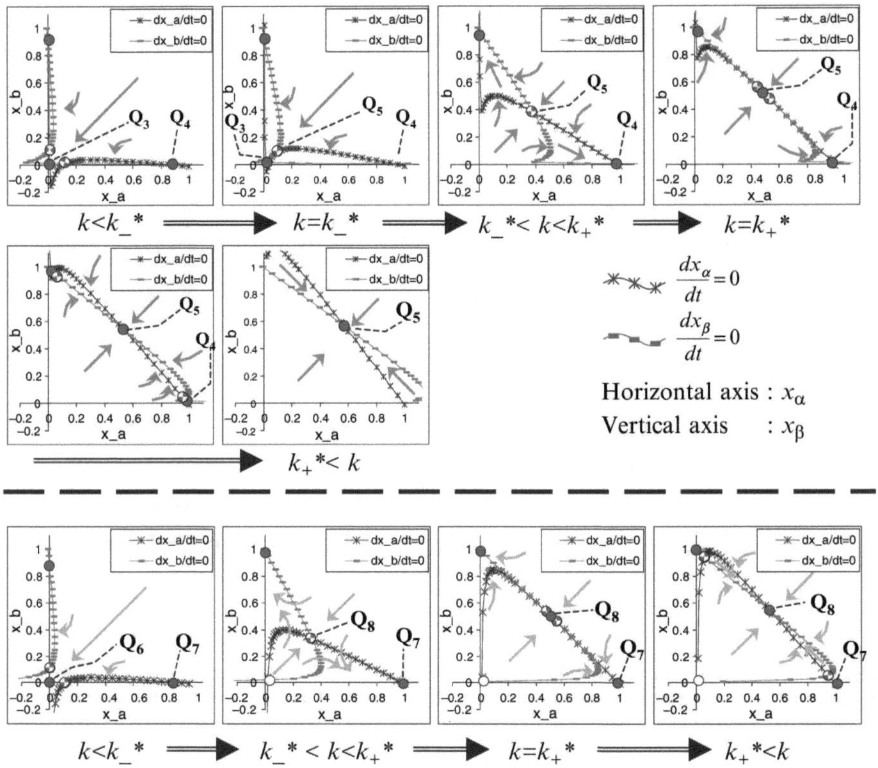

Fig. 5 Bifurcations of first-second dynamics (3) on the case of "isoclines with minima and maxima" (*above*) and "unimodal (*concave*) isoclines" (*below*)

5 Simulation Experimentation

Here we show simulation experiments of the emergence of money by the agent-based model without mean-field approximation. The first simulation demonstrates two analytical results from the previous section i.e. proto money can emerge even from homogeneous commodities, and no-emergence or single/multiple emergence of the proto-money depends on the degree of social network k. Changing the degree k of the regular-social network (250 agents), we conduct the simulation 200 times each, and observe the ratio of the number of emergence of proto-money from 32 homogeneous commodities.

Figure 6 shows the experimental result. The horizontal axis and the vertical axis express the network degree and the ratio of the number of each emerged proto-money. In Fig. 6, when the social network degree is small, no proto-money emerges. However, with increasing social network degree, single-emergence and multiple-emergence become dominant in turn. This observation supports our analytical results in the previous section.

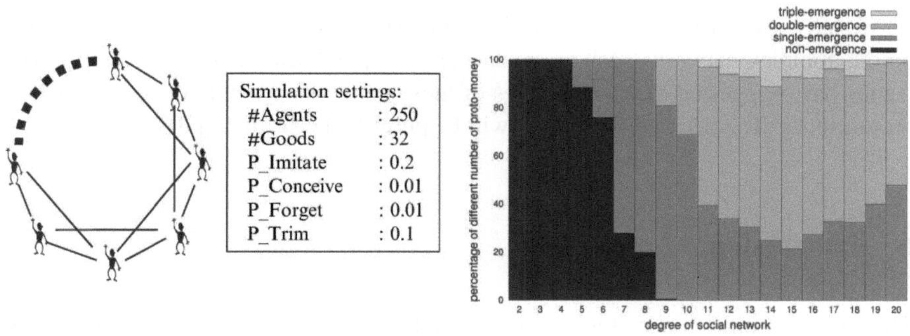

Fig. 6 The simulation on a regular social network illustrates that the network degree k (*horizontal axis*) determines the percentage of the emerged proto-money (non: *black*, single: *dark gray*, double: *gray*, more: *white*)

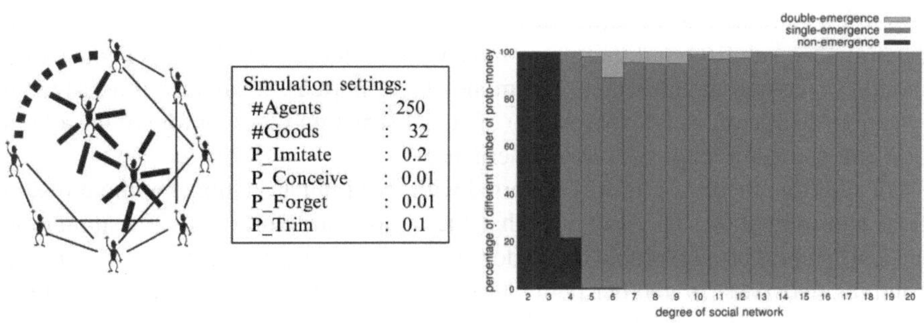

Fig. 7 Adding hub agents to previous simulation enforces the emergence of proto-money

The next experimentation shows a part of our work that illustrates a nature of the social network affects on the emergence of money. Literatures on complex network and epidemiology point out that the existence of hub nodes is an important factor for disease propagation. Similarly we can also expect a hub-effect in our model of the emergence of money.

A simple hub-effect will be observed by adding hub agents into the regular social network. Figure 7 shows an outcome on the modified regular social network by hub agents. Except for the 5% existence of hub agents (the horizontal axis represents average degree of social network), the other simulation parameter values are the same as previous experimentation (Fig. 6). Compared with Fig. 6, a modified social network with hub agents shows the emergence of money in lower average degree and the single emergence is more dominant than in the regular social network.

This hub-effect shows that some people who have high centrality of exchange of goods have an important role in the emergence of money. We often call such hub people of goods-exchange "merchants". Therefore the network analysis of the emergence of money implies the important existence of merchants.

We have also conducted intensive simulation studies on in the emergence of money on the complex networks such as small-world [27] and scale-free [2] networks. The results have suggested the complexities of those networks promote the emergence of proto-money. The detailed discussion will be published elsewhere [10].

6 Conclusion

This paper proposes the doubly structural network model of the emergence of money. This model enables us to easily describe a macroscopic emergence/self-organization phenomenon of proto-money from microscopic propagation/learning processes. As an analysis of the emergence of money, we build mean-field dynamics and first-second dynamics and carry out bifurcation analysis using the structural instability of the dynamics.

As a result, we can illustrate the emergence of proto-money (establishment of general acceptability) and that proto-money can emerge from homogeneous commodities (the non-metallic theory of money is supported). It can also illustrate that a nature (degree of social network: k) of social structure plays an important role in non-emergence or single/multiple emergence.

In addition, these results are supported with the simulation model without mean-field approximation. This suggests that the doubly structural network simulation analysis should be a favorable methodology for the emergence of money on various complex networks. In another publication, we will show interesting results via the doubly structured network simulation model on several types of complex social network.

The proposed DNS model of the emergence of money is applicable not only to the emergence of proto-money with hub agents discussed here but also to the emergence of virtual/private quasi-money. We extend our methods to the emergence of commonly exchangeable major mileage-points, which will be available in the literature [11].

Insight from this analysis on the emergence of proto-money as a transaction media suggests that the doubly structural network model can be an applicable method to other research on various types of communication in multi-agent societies.

References

1. Axelrod R (1997) The dissemination of culture: a model with local convergence and global polarization. J Conflict Resolut 41:203–326
2. Barabasi AL, Albert R (1999) Emergence of scaling in random networks. Science 286:509–512
3. Duffy J (2001) Learning to speculate: experiments with artificial and real agents. J Econ Dynam Contr 25:295–319
4. Epstein JS, Axtell R (1996) Growing artificial societies. The Brookings Inst, Washington, DC, ch. II–IV

5. Hayek FA (1976) Denationalization of money – the argument refined. The Institute of Economic Affairs, ch. 10/12
6. Hicks J (1967) Critical essays in monetary theory. Clarendon, Oxford, ch. 1
7. Iwai K (1996) The bootstrap theory of money: a search-theoretic foundation of monetary economics. Struct Change Econ Dynam 7:451–477
8. Kashima Y, Woolcock J, Kashima ES (2000) Group impressions as dynamic configurations: the tensor product model of group impression formation and change. Psychol Rev 107(4):914–941
9. Kiyotaki N, Wright R (1989) On money as a medium of exchange. J Polit Econ 97:927–954
10. Kobayashi M, Kunigami M, Yamadera S, Yamada T, Terano T (2009) Simulation modeling of emergence-of-money phenomenon by doubly structural network. Intelligent Decision Technologies
11. Kobayashi M, Kunigami M, Yamadera S, Yamada T, Terano T (2009) How a major mileage point emerges through agent interactions using doubly structural network model. In: WEIN'09 (Workshop on Emergent Intelligence on Networked Agents)
12. Klemm K, Eguiluz VM, Toral R, Miguel MS (2003) Nonequilibrium transitions in complex networks: a model of social interaction. Physical Review E 67:026120-1–026120-6
13. Kunigami M, Kobayashi M, Yamadera S, Terano T (2007) On emergence of money in self-organizing micro-macro network model. In: Proceedings of ESSA'07, pp 417–425
14. Luo GY (1999) The evolution of money as a medium of exchange. J Econ Dynam Contr 23:415–458
15. Mankiw NG (1999) Macroeconomics, 4th edn. Worth Pub, New York, ch. 5
16. Marimon R, McGrattan E, Sargent TJ (1990) Money as a medium of exchange in an economy with artificially intelligent agents. J Econ Dynam Contr 14(2):329–373
17. Masuda N, Konno N (2006) Multi-state epidemic processes on complex networks. J Theor Biol 243:64–75
18. Menger C (1871) Grandsatze der Vorkswwirtschaftslehre, ch. 8
19. Menger C (1923) Grandsatze der Vorkswwirtschaftslehre (Zweite Aufage), ch. 9
20. Myerson AR (1990) Business diary. The New York Times
21. Pastor-Satorras R, Vespignani A (2002) Immunization of complex networks. Phys Rev E 65:36104–36118
22. Quiggin AK (1949) A survey of primitive money – the beginnings of currency. Methuen & Co., London, pp 25–318
23. Radford RA (1945) The economic organisation of a P.O.W. camp. Economica 12:189–201
24. Reynolds A (1993) Monetary reform in Russia: the case for gold. Cato Journal 12(3): 657–676, p 667
25. Rostworowski De Diez Canseco M (1988) Historia del Tahuantinsuyu. Instituto de Estudios Peruanos, Lima
26. Starr RM (2003) Why is there money? Endogenous derivation of 'money' as the most liquid asset: a class of examples. Econ Theor 21:455–474
27. Watts DJ, Strogatz SH (1998) Collective dynamics of small-world networks. Nature 393:1302–1305
28. Weber M (1920) Die protestantische Ethik und der 'Geist' des Kapitalismus
29. Yamadera S, Terano T (2007) Examining the myth of money with agent-based modeling. In: Edmonds B et al (eds) Social simulation – technologies, advances, and new discoveries. Information Science Reference, Hershey, pp 252–262
30. Yasutomi A (1995) The emergence and collapse of money. Physica D 82:180–194

Analysis of Knowledge Retrieval Heuristics in Concurrent Software Development Teams

Shinsuke Sakuma, Yusuke Goto, and Shingo Takahashi

Abstract This paper examines effective knowledge-retrieval heuristics in terms of transactive memory in concurrent software development teams. We propose six knowledge-retrieval heuristics and evaluate their effectiveness in four typical situations encountered by concurrent software development teams. Agent-based social simulation suggests the following findings about effective heuristics in each situation: (1) in large teams, if team members have incomplete information about their teammates' expertise, both the minimum effort type and the risk aversion type are effective; (2) in small teams, if team members have incomplete information about their teammates' expertise, the broad retrieval type is effective; (3) in small teams with a heavy work-load, if team members have sufficient information about their teammates' expertise, the minimum effort type is effective; (4) in small teams with a light work-load, if team members have sufficient information about their teammates' expertise, both the minimum effort type and the "ask others" type are effective.

Keywords Knowledge retrieval • Concurrent process • Transactive memory • Organizational learning

S.S akuma
NomuraR esearchI nstitute, 1-6-5M arunouchi, Chiyoda,T okyo100- 0005,J apan
e-mail:s -sakuma@nri.co.jp

Y.G oto(✉)
IwatePr efecturalU niversity, 152-52S ugo, Takizawa, Iwate020- 0193,J apan
e-mail:y- goto@iwate-pu.ac.jp

S.T akahashi
WasedaU niversity, 3-4-1O kubo, Shinjuku, Tokyo 169-8555, Japan
e-mail:s hingo@waseda.jp

K. Takadama et al. (eds.), *Simulating Interacting Agents and Social Phenomena: The Second World Congress*, Agent-Based Social Systems 7,
DOI 10.1007/978-4-431-99781-8_11, © Springer 2010

1 Introduction

Although rapid software development has always been in demand; in recent times, that demand has intensified. Most software development teams are overhauled for each new project, based on its requirements and the availability of team members [2]. The members of such a software development team perform assigned tasks by applying their knowledge, which is usually specialized in their respective fields. Team members concurrently perform their assigned tasks and ultimately produce the components they are individually responsible for. The team's output is composed of these individual components that are developed and integrated by the team.

One problem encountered by concurrent software development teams is the effective sharing of knowledge among team members. According to an investigation of 69 concurrent software development teams carried out by Faraj and Sproull [2], the metaknowledge of "who knows what" and "what is known by whom" plays an important role in determining a team's performance. Wegner and Wegner [8] defined transactive memory (TM) as a shared system for encoding, storing, and retrieving information from different domains, developed by people working in closely knit teams. Knowing "who is good at what" enables members to effectively retrieve the required knowledge which results in an improvement in the team's performance.

While most laboratory experiments and agent-based social simulations suggest that knowledge retrieval, or knowledge recall, that uses TM has a positive impact on effective knowledge-sharing within a team [3,5,7], a negative impact of collaborative inhibition due to retrieval disruption has also been noted [1]. It is therefore necessary to ascertain the effective heuristics of knowledge retrieval in terms of TM, and to build robust TM systems into the team.

The purpose of our research is to examine effective heuristics of knowledge retrieval in terms of TM in concurrent software development teams. We propose six knowledge retrieval heuristics and evaluate their effectiveness in four typical situations encountered by concurrent software development teams.

2 Model

2.1 PCANNS Scheme

We have modeled a concurrent software development team as a multiagent system by using the PCANNS scheme [7], which specifies three elements people (*P*), resources (*R*), and tasks (*T*) and the following six relationships:

1. **P**recedenceoft asks($T \times T$)
2. **C**apabilities linking people to resources ($P \times R$)
3. **A**ssignment of tasks to people ($P \times T$)
4. **N**etworksa mongpe ople($P \times P$)

5. Resource Needsofta sks($R \times T$)
6. Substitutesofre sources($R \times R$)

A team is represented as six relational matrices in which elements are either 1 or 0 (1 signifies there is a connection between two elements; 0 signifies there is no connection).

2.2 Concurrent Software Development Team

Each agent in the software development team is assigned a series of tasks that he/she resolves concurrently. $P_{jj'}$ represents the precedence of task j over j' ($j, j' = 1, 2, \cdots, TS$). Let TS be the number of tasks. In the team, the elements of the $T \times T$ matrix $P_{jj'}$ are initially both defined and fixed. Let AS be the number of agents in the team. A_{ij} represents whether agent i ($= 1, 2, \cdots, AS$) is assigned to task j. The elements of the $P \times T$ matrix A_{ij} are also initially defined and fixed. Let AT_i be the number of tasks assigned to agent i. Tasks are assigned to agents in such a way that equal weight is given to all agents ($AT_i = TS \ / AS$). Agent i is assigned to task j ($= (i-1) \times AT_i + 1, \cdots, i \times AT_i$).

Each task has a deadline. Tasks are performed only until their respective deadlines are met, even if they are not fully accomplished by that time. The outputs of the tasks performed by each agent are then combined into an integrated organizational output (Fig. 1).

Agents share close working relationships. $NT_{ii'}$ represents whether agent i can communicate with agent i' ($= 1, 2, \cdots, AS$). The elements of the $P \times P$ matrix $NT_{ii'}$ are always 1. All agents can communicate with each other.

Some units of knowledge are required to perform a given task. Let KS be the number of domains from which knowledge is utilized. N_{kj} represents whether knowledge k ($= 1, 2, \cdots, KS$) is required to solve task j. The elements of the $R \times T$ matrix N_{kj} are initially determined randomly and are fixed. $S_{kk'}$ represents whether knowledge k can be substituted for knowledge k' ($= 1, 2, \cdots, KS$). The elements of the $R \times R$ matrix $S_{kk'}$ are always 0 because we do not consider the substitution of knowledge.

Agents have some suitable level of knowledge to perform their assigned tasks. The elements C_{ik} denote whether or not agent i has knowledge k. The elements

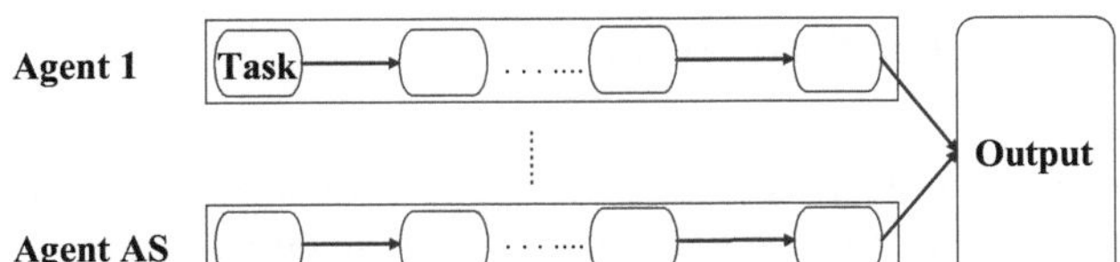

Fig. 1 Concurrent task process

of the $P \times R$ matrix $C_{ik} = \{0,1\}$ are initially set but change through the acquisition of knowledge during knowledge retrieval.

The agents refer to their own TM for "who knows what," and retrieve the required information by communicating with other agents in the team. Their TM is modified as a result of such communication. Let ${}_iC_{i'k}$ denote agent i's expectation regarding whether agent i' has knowledge k. The element ${}_iC_{i'k}$ is represented as a trinary (${}_iC_{i'k} = \{-1,0,1\}$):

$$
{}_iC_{i'k} = \begin{cases} -1 & \text{if } i \text{ deems that } i' \text{ does not have } k, \\ 0 & \text{if } i \text{ lucks information about whether } i' \text{ has } k \text{ or does not,} \\ 1 & \text{otherwise.} \end{cases}
$$

All agents have their own $P \times R$ matrices as their TM, and the element of the matrix ${}_iC_{i'k}$ can be misrecognized. As an exception, if $i = i'$, ${}_iC_{i'k}$ ispre ciselyre cognized. The value of element ${}_iC_{i'k}$ changes as a result of the communications that take place during knowledge retrieval.

In conventional software development teams, if the initial work is of poor quality, it is hard to achieve the desired quality in the subsequent stages of work. Therefore, in this model, any given task's accomplishment is affected by the output of the preceding task. Let $P(i,j')$ be the performance of task j (j'th task assigned to agent i):

$$
P(i,j') = \begin{cases} \dfrac{\sum_k |N_{kj} \times C_{ik}|}{\sum_k N_{kj}} & \text{if } j' = 1, \\[2ex] \sqrt{P(i,j'-1)} \times \dfrac{\sum_k |N_{kj} \times C_{ik}|}{\sum_k N_{kj}} & \text{otherwise.} \end{cases}
$$

In concurrent software development teams, the team's output is composed of components that are implemented and integrated by the members. Therefore, the organizational performance would be the average of the individuals' task resolution performance. We define the organizational performance OP as the average of $P(i,j')$:

$$
OP = \sum_i \sum_{j'} \frac{P(i,j')}{AS}.
$$

2.3 Knowledge Retrieval Heuristics

Agent i assigned to task j retrieves knowledge k when it is essential to perform task j ($N_{kj} = 1$), subject to the condition that agent i does not have the requisite knowledge k ($C_{ik} = 0$). Agents acquire the knowledge they need on their own or acquire that knowledge through knowledge retrieval communications. Using knowledge retrieval heuristics, agents determine "whom" to ask for "what." We classify what agent i asks in the following four patterns: (1) the knowledge k being retrieved; (2) the queried agent i''s knowledge ($C_{i'1},\cdots,C_{i'KS}$); (3) the queried agent i''s TM about who has the knowledge being retrieved (${}_{i'}C_{1k},\cdots,{}_{i'}C_{ASk}$); and (4) the queried agent's TM about who knows what (${}_{i'}C_{11},\cdots,{}_{i'}C_{ASKS}$). We classify the agents from whom agent i requests knowledge into three types: (A) agent i', who seems to have the most knowledge ($i' = \arg\max \sum_{k} {}_{i}C_{i'k}$); (B) agent i', who seems to have the knowledge k being retrieved (${}_{i}C_{i'k} = 1$); (C) agent i', who is randomly selected. These classifications are defined from a normative rather than realistic viewpoint, and are referred to in the description of the knowledge-retrieval heuristics.

We introduce the following six types of knowledge-retrieval heuristics. See Appendix for detailed algorithms for each of them.

1. *Minimum effort type*. Agent i requests (1) from (A). In this heuristic, the agent uses his/her TM to obtain the required knowledge from other agents. This heuristic tends to minimize his/her effort.
2. *Risk aversion type*. Agent i requests (2) from (A); subsequently, i requests (1) from agent (A). In this heuristic, the agent uses his/her TM to develop his/her knowledge from other agents, and to retrieve the required knowledge from team members. This heuristic tends to averse knowledge-retrieval failures.
3. "*Ask others*" *type*. Agent i requests (1) from (A), then i requests (1) from (B). Moreover, i requests (3) from (A). In this heuristic, the agent not only uses his/her TM to request knowledge from other agents, but also develops his/her TM to retrieve that knowledge with certainty. This heuristic tends to encourage agents to ask others.
4. "*Acquire on my own*" *type*. Agent i tries to acquire the required knowledge k on his/her own. In this heuristic, the agent does not use his/her TM to retrieve knowledge. Therefore, this heuristic serves as a benchmark for the other heuristics.
5. *Broad retrieval type*. Agent i requests (4) from (A), then i requests (1) from (A). In this heuristic, the agent uses his/her TM to develop his/her TM as well as efficiently retrieve the required knowledge from other agents. This heuristic tends to encourage broad knowledge-retrieval among team members.
6. *Random type*. Agent i requests (1) from (C) or tries to acquire k on his/her own. In this heuristic, the agent does not use his/her TM to retrieve knowledge. Therefore, this heuristic serves as a benchmark for the other heuristics.

2.4 Simulation Flow

Figure 2 describes a flowchart of the simulation. Agent i retrieves domain knowledge k with one of the six types of knowledge-retrieval heuristics. Knowledge retrieval entails varying amounts of time spent that depend upon "what" is asked of "whom." Table 1 shows an example of time spent in knowledge retrieval ($AS = 9$). In each setting, time spent is estimated on the basis of the results obtained by Inuzuka and Nakamori's experiment [4]. Agent i performs assigned task j when he/she has enough knowledge to perform the task or the deadline is reached.

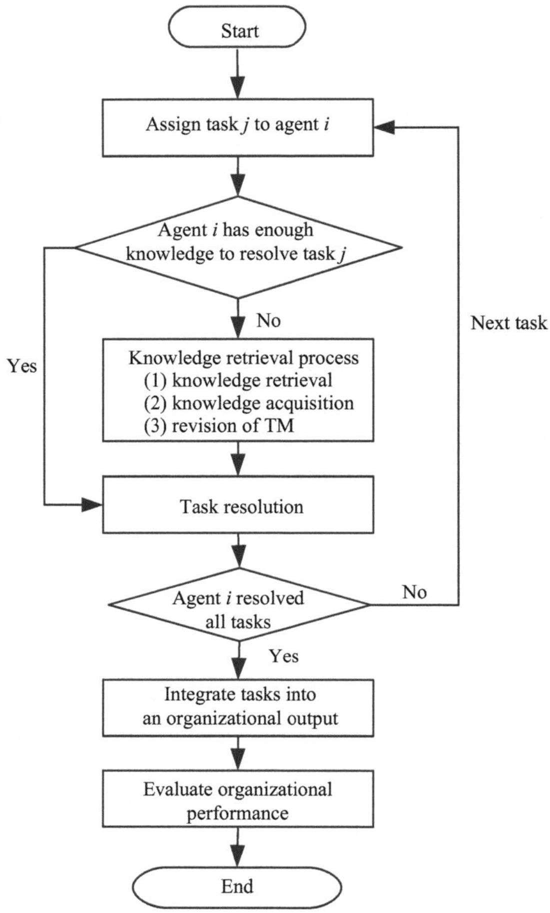

Fig. 2 Simulation flowchart

Table 1 Example of time cost in knowledge retrieval (AS = 9)

	Knowledge retrieval in terms of TM	Steps to be accomplished
What	(1) The knowledge being retrieved	971
	(2) Queried agent's knowledge	300
	(3) Queried agent's TM about who has the knowledge being retrieved	9
	(4) Queried agent's TM about "who" knows "what"	2,700
To whom	(A) The agent who seems to have the most knowledge	300
	(B) The agent who seems to have knowledge being retrieved	609
	(C) A randomly selected agent	0

3 Parameter Calibration

Ren et al. [7] calibrated their model using data from a laboratory experiment. They suggested that the use of TM helps smaller teams make better quality decisions more than it helps larger teams. In their model, "quality" corresponds to "organizational performance" in our model. In this section, we estimate whether selected parameters in our model are consistent with their suggestion.

We divided the six knowledge-retrieval heuristics into two groups: heuristics that use TM and heuristics that do not. The former includes the *minimum effort type*, *risk aversion type*, *"ask others" type*, and *broad retrieval type*. The latter includes the *"acquire on my own" type* and *random type*. We introduce two indices for TM, the density of agent i's TM Dn_i and the accuracy of agent i's TM Ac_i:

$$Dn_i = \frac{\sum_{i'}\sum_{k}\left|{}_iC_{i'k}\right|}{KS \times AS},$$

$$Ac_i = \frac{\sum_{i'}\sum_{k} e_{i'k}}{\sum_{i'}\sum_{k}\left|C_{i'k}\right|},$$

$$e_{i'k} = \begin{cases} 1 & \text{if } ({}_iC_{i'k} = \mathrm{C}_{i'k} = 1) \text{ or } ({}_iC_{i'k} = -1, \mathrm{C}_{i'k} = 0), \\ 0 & \text{otherwise.} \end{cases}$$

In the calibration process, initially we set $Dn_i = 0.5$, $Ac_i = 0.5$, the number of knowledge domain $KS = 500$, and the number of tasks assigned to agent i $AT_i = 14$. There are 120,000 time steps before the deadline of each task. This setting assumes a balanced, common situation in which the parameter values seem to be average-level ones.

We calibrated the following three parameters: the probability that the knowledge requested is successfully transmitted R_T, the probability that the required knowledge is successfully acquired by the agent on his/her own R_A, and the probability that the agent acquires the required knowledge from a randomly selected agent R_S. After the calibration, we set $R_T = 0.9$, $R_A = 0.3$, and $R_S = 0.5$. Figure 3 shows that the result obtained by the calibrated model is consistent with the suggestion made by Ren et al. [7].

4 Effective Knowledge Retrieval Heuristics

In this section, we evaluate the effectiveness of the six proposed knowledge-retrieval heuristics. In this paper, we show the results in four typical situations (see Table 2). In all situations, 30 units of knowledge are required for task resolution; each agent has 30 units of knowledge; 80,000 steps are carried out before the deadline of

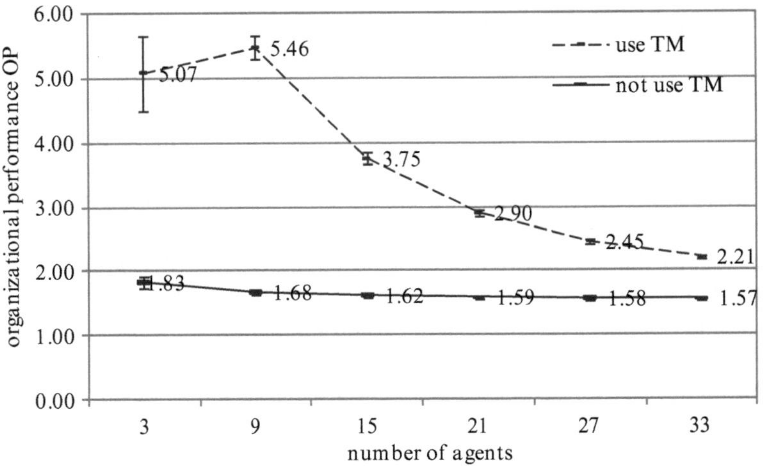

Fig. 3 Results of the experiment obtained by using the calibrated parameters

Table 2 Experimental design

	Experiment			
	1	2	3	4
Initial TM density	0.1	0.1	0.9	0.9
Initial TM accuracy	0.1	0.1	0.9	0.9
Number of agents	33	9	9	9
Number of the assigned tasks	18	10	18	10

each task. The average performance of 100 runs of the knowledge-retrieval heuristics are shown in Figs. 4–7. We randomly changed the agents' knowledge, TM, and tasks in each run. In all the experiments, the differences between the performance of the heuristics deemed effective and those deemed ineffective are statistically significant.

Experiment 1 assumes that the team is large and its agents do not know each other. Figure 4 suggests that the *minimum effort type* and the *risk aversion type* are effective. Experiment 2 assumes that the team is small and its agents do not know each other. Figure 5 suggests that the *broad retrieval type* is most effective. If agents' TM is poor, the development of TM is an apparent prerequisite for effective knowledge retrieval. For a large team, however, the cost of such development could exceed its value. On such a large team, even one that performs poorly, the utilization of TM better contributes to organizational performance than the development of TM.

Experiment 3 assumes that the team is small and with a heavy load, and its agents know each other well. Figure 6 suggests that the *minimum effort type* is most effective. Experiment 4 assumes that the team is small, and its agents know each other well. Figure 7 suggests that the "*ask others*" *type* and the *minimum effort type* are effective.

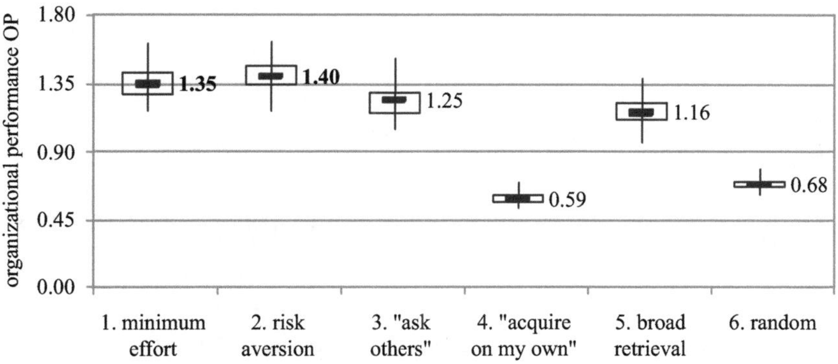

Fig. 4 Experiment 1: large team and poor TM

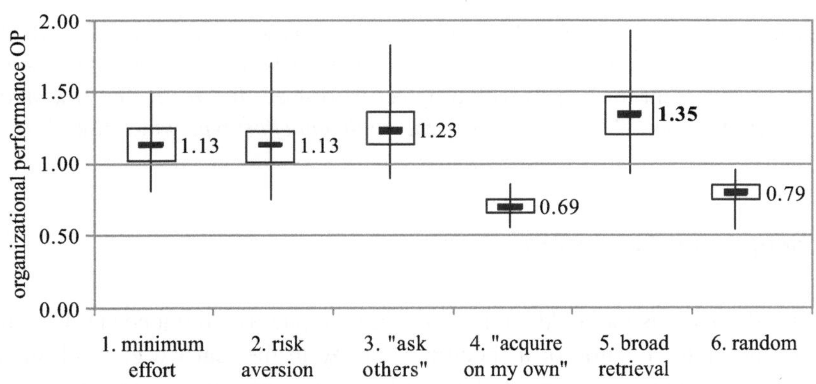

Fig. 5 Experiment 2: small team and poor TM

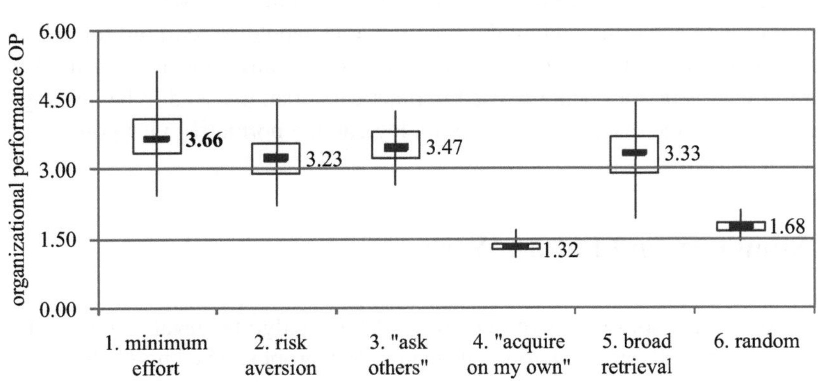

Fig. 6 Experiment 3: small team and rich TM with a heavy load

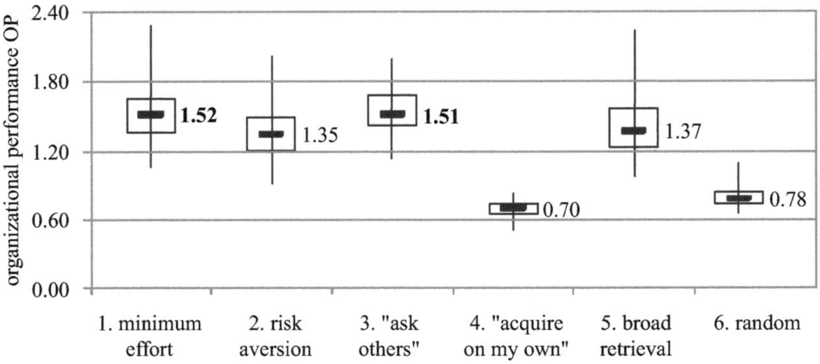

Fig. 7 Experiment 4: small team and rich TM.

If agents' TM is rich, the development of TM seems to be necessary for effective knowledge-retrieval. The effect of such development, however, could compete with the cost of development for a team with a lighter work-load. In Fig. 7, the "*ask others*" *type* which develops TM competes with *the minimum effort type* which uses TM.

5 Scope and Grounding

Our ABSS model is the middle range one, not the one which intends to describe a specific problem situation of a specific company in the real world. Such middle range models aim to describe the characteristics of a particular social phenomenon in a sufficiently general way that their findings can be applied widely [6]. We think that our findings have sufficient applicability to such situations in which tasks have their deadline and are performed concurrently by a team of intellectual workers which needs effective communication to retrieve required knowledge.

Further data gathering and model specifications would be required for, when our model intends to ground on a specific real-world company. For instance, the knowledge-retrieval heuristics actually used in practical situations need to be investigated, because we use the ones proposed from a logical and normative viewpoint.

6 Summary and Future Work

This paper establishes effective heuristics of knowledge retrieval in terms of TM, for concurrent software development teams. We propose six knowledge-retrieval heuristics and evaluate their effectiveness in four typical situations encountered by concurrent software development teams.

ABSS suggests the following findings about effective heuristics in each situation: (1) in large teams, if team members have incomplete information about their teammates' expertise, both the minimum effort type and the risk aversion type are effective; (2) in small teams, if team members have incomplete information about their teammates' expertise, the broad retrieval type is effective; (3) in small teams with a heavy work-load, if team members have sufficient information about their teammates' expertise, the minimum effort type is effective; (4) in small teams with a light work-load, if team members have sufficient information about their teammates' expertise, both the minimum effort type and the "ask others" type are effective. These findings would contribute to more effective communication in these teams.

A future task would be to analyze these results in more depth. Time-series log analysis that focuses on the micro-level behavior of knowledge-retrieval would deepen our understanding of why and how a specific heuristic promote better performance.

Acknowledgments This work was supported in part by a Grant-in-Aid for Scientific Research 21310097 of JSPS and a Grant for Special Research Projects of Waseda University (2009B-176).

Appendix: Algorithm of Knowledge-Retrieval Heauristics

Minimum effort type

1. Agent i requests the required knowledge k from the agent $i'(i' = \arg\max \sum_{k'\ i} C_{i'k'})$. Go to 2
2. Resolveta sk j? (Yes => Task resolution; No => Go to 3)
3. i acquired k? (Yes => Go to 1; No => Go to 4)
4. i requests k from i' suchtha t ${}_i C_{i'k} = 1$ .G oto5
5. Resolve j ? (Yes => Task resolution; No => Go to 6)
6. i acquired k? (Yes => Go to 4; No => Go to 7)
7. Execute *random type*

Risk aversion type

1. Agent i requestsa gent i''sTM $C_{i'1}, \cdots, C_{i'KS}$ $(i' = \arg\max \sum_{k'\ i} C_{i'k'}$).G oto2
2. Resolve j? (Yes => Task resolution; No => Go to 3)
3. i requests k from $i'(i' = \arg\max \sum_{k'\ i} C_{i'k'}$).G oto4
4. Resolve j? (Yes => Task resolution; No => Go to 5)
5. i acquired k ? (Yes => Go to 6; No => Go to 7)
6. i' hadkno wledge k? (Yes => Go to 3; No => Go to 1)
7. i requestsa gent i''sTM $C_{i'1}, \cdots, C_{i'KS}$ from i' sucht hat ${}_i C_{i'k} = 1$ .G oto8

8. Resolve j? (Yes => Task resolution; No => Go to 9)
9. i requests k from i' such that ${}_iC_{i'k} = 1$. Go to 10
10. Resolve j? (Yes => Task resolution; No => Go to 11)
11. i acquired k? (Yes => Go to 12; No => Go to 13)
12. i' had knowledge k? (Yes => Go to 9; No => Go to 7)
13. Execute *random type*

"Ask others" type

1. Agent i requests the required knowledge k from agent i' ($i' = \arg\max \sum_{k'} {}_iC_{i'k'}$). Go to 2
2. Resolve task j? (Yes => Go to 12; No => Go to 5)
3. i acquired k? (Yes => Go to 1; No => Go to 4)
4. i requests k from i' such that ${}_iC_{i'k} = 1$. Go to 5
5. Resolve task j? (Yes => Go to 12; No => Go to 6)
6. i acquired k? (Yes => Go to 4; No => Go to 7)
7. i requests i''s TM ${}_{i'}C_{1k}, \cdots, {}_{i'}C_{ASk}$ from i' ($i' = \arg\max \sum_{k'} {}_iC_{i'k'}$). Go to 8
8. Resolve j? (Yes => Go to 12; No => Go to 9)
9. i requests k from i' such that ${}_iC_{i'k} = 1$. Go to 10
10. Resolve task j? (Yes => Go to 12; No => Go to 11)
11. i acquired k? (Yes => Go to 9; No => Go to 7)
12. Task resolution

"Acquire on my own" type

1. Agent i tries to acquire the required knowledge k on his/her own. Go to 2
2. Resolve j? (Yes => Go to 3; No => Go to 1)
3. Task Resolution

Broad retrieval type

1. Agent i requests i''s TM ${}_{i'}C_{11}, \cdots, {}_{i'}C_{ASKS}$ from agent i' ($i' = \arg\max \sum_{k'} {}_iC_{i'k'}$). Go to 2
2. Execute *minimum effort type*

Random type

1. (Atth eproba bilityof R_s =>G oto2; $1-R_s$ =>G oto3)
2. Agent i tries to acquire the required knowledge k on his/her own. Go to 4
3. i requests k froma gent i',s electedra ndomly.G oto4
4. Resolve j? (Yes => Task resolution; No => Go to 1)

References

1. Basden BH, Basden DR, Bryner S, Thomas RL III (1997) A comparison of group and individual remembering: does collaboration disrupt retrieval strategies? J Exp Psychol Learn Mem Cognit 23:1176–1189
2. Faraj S, Sproull L (2000) Coordinating expertise in software development teams. Manag Sci 46:1554–1568
3. Hollingshead AB (1998) Communication, learning, and retrieval in transactive memory systems. J Exp Soc Psychol 34:423–442
4. Inuzuka A, Nakamori Y (2003) A recommendation for IT-driven knowledge sharing. Trans Inst Electron Inform Commun Eng J86-D-I:179–187 (in Japanese)
5. Liang DW, Moreland R, Argote L (1995) Group versus individual training and group performance: the mediating role of transactive memory. Pers Soc Psychol Bull 21:384–393
6. Gilbert N (2007) Agent-based models. Sage Publications, London
7. Ren Y, Carley KM, Argote L (2006) The contingent effects of transactive memory: when is it more beneficial to know what others know? Manag Sci 52:671–682
8. Wegner TG, Wegner DM (1995) Transactive memory. In: Manstead ASR, Hewstone M (eds) The Blackwell encyclopedia of social psychology. Blackwell, Oxford

Reputation and Economic Performance in Industrial Districts: Modelling Social Complexity Through Multi-Agent Systems

Gennaro Di Tosto, Francesca Giardini, and Rosaria Conte

Abstract Industrial districts (Ids) can be conceived as complex systems made of heterogeneous but strictly interrelated and complementary firms that interact in a non-linear way. One of the distinctive features of industrial districts is the tight connection existing between the social community and the firms: in this context, economic exchanges are mainly informed by social relationships and holding good reputation is an asset that may actually foster potential relations. In this work we model the effects of social evaluations on firms in an artificial cluster through Multi-Agent Simulation (MAS) techniques, in order to investigate whether and how different kinds of social evaluations have an impact on firms' quality and on their profits. Likewise, we compare the effects of sincere and insincere information on the economic performances of the single firms and of the cluster as a whole.

Keywords Reputation • Partner selection • Economic behavior • Agent-based simulation

1 Introduction

Social evaluations are pieces of information regarding other agents whose attitudes, behaviours and actions are assessed with respect to some specific dimensions or aspects. Individuals use these evaluations as guidance to predict others' behaviours and to choose the most appropriate response when first-hand experience is not available or when it is too costly, in terms of risk, time and energy, to be acquired.

G. Di Tosto (✉), F. Giardini, and R. Conte
Instuto di Scienze e Tecnologie della Congnizione, CNR, Via San Martino della, Battaglia 44, 00185,R omaI taly
e-mail: gennaro.ditosto@istc.cnr. it; francesca.giardini@istc.cnr. it; rosaria.conte@istc.cnr.it

K. Takadama et al. (eds.), *Simulating Interacting Agents and Social Phenomena: The Second World Congress*, Agent-Based Social Systems 7,
DOI 10.1007/978-4-431-99781-8_12, © Springer 2010

Transmitting social evaluations, i.e. gossiping, is crucial in human societies, in which gossip facilitates the formation of groups [10]: gossipers share and transmit relevant social information about group members within the group, at the same time isolating out-groups. Besides, gossip contributes to stratification and social control, since it works as a tool for sanctioning deviant behaviours and for promoting, even through learning, those behaviours that are functional with respect to the group's goals and objectives, mainly norms and institutions [22]. Sommerfeld et al. [19] consider gossip as a way to transfer social information within groups, alternative to direct observation. This flow of information maintains cooperation by indirect reciprocity.

Furthermore, reputation is considered pivotal in creating and sustaining prosocial behaviours in large human groups. Theories of indirect reciprocity show how cooperation in large groups can emerge when the agents are endowed with, or can build, a reputation [2, 9, 16]. According to this theory, large scale human cooperation can be explained in terms of conditional helping by individuals who want to uphold a reputation and then to be included in future cooperation [17], as demonstrated by several experimental studies (for an introduction, see [8]). Reputational information can also solve the *tragedy of the commons*, a social dilemma referring to the fact that a public good will be overused if everybody is allowed to do so [21]. Allowing people to build up a reputation, prevented the public resource from being overused.

Although influential, these theories suffer the flaw of what Granovetter [11] calls an *undersocialized* notion of reputation.

> Economists have pointed out that one incentive not to cheat is the cost of damage to one's reputation; but this is an undersocialized conception of reputation as a generalized commodity, a ratio of cheating to opportunities for doing so. In practice, we settle for such generalized information when nothing better is available [11, p. 490].

Granovetter points out the relevance of information coming from one's own past dealings with someone, highlighting the benefits of this second kind of information that is cheap, more detailed, and, of course, accurate. This kind of information can be easily acquired thanks to *embeddedness*, i.e. the fact that human actions are motivated and explained by their being embedded in a network of social relationships that foster cooperation and guarantee against cheaters.

Evolutionary models have demonstrated how cooperation in large groups or without repeated interaction can be sustained if individuals' payoffs are reduced by their reputation as bad contributors to public goods [12, 13]. In other words, reputation is a cheap solution to solve cooperative dilemmas in which individuals are required to pay a cost to create a benefit for the group. This solution has been adopted also in electronic commerce websites, in which users can evaluate their peers, giving feedbacks that future users can use to avoid frauds and select the best peers [7].

Social evaluations become critical in closed environments in which the web of relationships among agents determines their behaviours, actions and results. This happens, for instance, in industrial districts,[1] in which the interplay between

[1] In this work, *cluster* and *district* are used as synonyms.

economic dimensions and social relationships is very close. Industrial clusters are usually defined as networks of interactions among heterogeneous and complementary firms embedded into a specific geographic area. In the district, the form of production requires a high degree of cooperation between firms and the lack of formal agreements could lead actors to behave in an opportunistic manner, but the merging between social community and firms [3] helps preventing this result.

The close interplay between social and economic aspects that characterizes industrial districts has drew the attention of several scholars aimed at building up sound models of clusters' interactions and economic performances. In this work we will focus upon modelling and simulation of artificial districts, without taking into account research coming from organization and management studies. In order to manage the complexity of this phenomenon, two different strands of research have been followed. According to the *realistic* approach, detailed and realistic models of districts' processes and results are required to understand the phenomenon [4, 20]. On the other hand, D. Lane [15] proposes an approach in which the lack of realism is counterbalanced with the introduction of cognitive and social factors to explain the cluster's complexity. In this work we will follow this second strand and we will model an artificial industrial district moving from the definition given by Albino, Carbonara, and Giannoccaro [1]: "IDs are geographically defined productive systems characterized by a large number of small and medium sized firms that are involved at various phases, and in various ways, in the production of a homogeneous product. These firms are highly specialized in a few phases of the production process, and integrated through a complex network of inter-organizational relationships. A close relationship between the social, political, and economic spheres further characterizes the IDs".

1.1 Research Hypothesis

The aim of this work is to couple the model of an artificial cluster with a cognitive account of social evaluations. Reputation is a cognitive and social artifact rooted in individual minds but acting at the supra-individual level and evolved to solve collective problems. We adopt the socio-cognitive framework developed by [6], who describe how people create, manipulate and transmit social evaluations, and how these evaluations affect individuals' beliefs and behaviours. This approach is a dynamic one that considers reputation as the output of a social process that starts in agents' minds. Notably, this theory applies not only to humans, but also to artificial agents in a variety of distinct environments [18]. According to the theory, the evaluation that agents (Evaluators) directly form about a given Target during interaction or observation represents the input of the process. This evaluation can be transmitted to Beneficiaries that share the goal with regard to which targets are evaluated and thus may use this information as a guide for their behaviour: knowing in advance others' behaviours and attitudes may thwart cheaters. The social and cognitive account of reputation proposed here allows:

1. To distinguish between *image*, the output of a process of evaluation in which is made explicit who made that evaluation, and *reputation*, in which the source is impersonal. Image and reputation are both social evaluations about a Target, but they differ with regard to the identity of the source. Evaluations from a nameless source can be less reliable, but they do not expose the gossiper to retaliation, as it happens with image.
2. To account for the cognitive determinants of reputation and for its dynamic effects, both at the individual and at the collective level.
3. To predict the agents' behaviours and resulting actions at the macro-level.

The social and cognitive theory of reputation clearly distinguishes between two classes of social evaluations, image and reputation, and it has been also showed that they affect the survival rate of two different populations of cheaters and cooperators, enforcing norm conformity [5]. The aim of this paper is to test whether social evaluations may produce an effect in a complex scenario, characterized by close relationships among agents. At present, the model is really basic and it lacks both evolutionary processes and a truly cognitive agents' architecture but, as far as we know, this is the only model in which agents can communicate and use social evaluations to orient their choices.

In what follows, we will present a simulation model of reputation and its transmission in an artificial cluster of firms. The relevance of social evaluations in this context makes it suitable to verify the socio-cognitive theory of reputation, and to test whether and in what way the exchange of social information can be related to the quality of products delivered by artificial firms. The application of an agent-based computational approach to the study of industrial districts is not new (for a review see [14]), but we want to add to this literature by using cognitive agents that manipulate and circulate two different kinds of social evaluations, i.e. image and reputation.

Social and cognitive effects are implemented through communication processes. Reputation and gossip are responsible for the changes in the economic performances monitored throughout the simulations. The experimental work described will report on the changes in quality and profit of the production chains as an effect of the types of evaluation transmitted and the strategy of interaction between firms during the communication process.

The stylized nature of the processes implemented, even though it can present some limitation on the economic side, allows us to test the presence of a causation link between the cognitive variables of the model and the economic outcome, taking into account the cognitive complexity of the mechanism designed.

2 The Simulation Model

The agents of the model are firms. They all produce components that are assembled into the only kind of final product sold in the market. Their goal is to select the best available supplier (in terms of goods' quality) in order to maximize their profits,

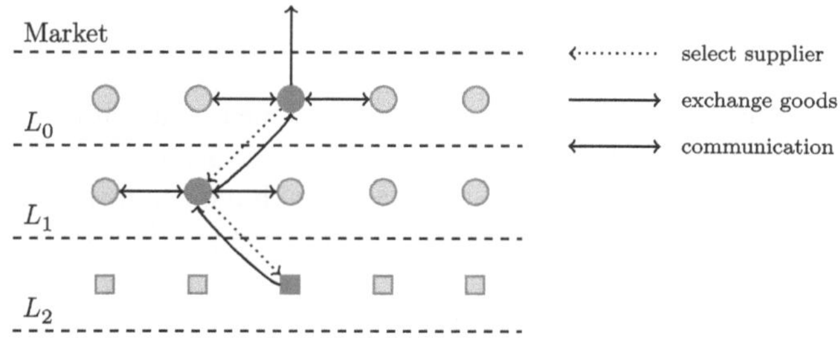

Fig. 1 Agents interactions: firms select suppliers from the lower level, which in turn provide them components for their products, and communicate with firms belonging to the same level, exchanging evaluations about suppliers. L_2 firms are producers of raw materials and they do not communicate with each other; they are only chosen as suppliers by the firms of the layer above them

and similar firms may collaborate with each other exchanging evaluations about known, tested suppliers (see Fig. 1).

2.1 *Partner Selection and Economic Exchange*

Firms are organised into different layers, with each layer containing agents that act as suppliers for the firms of the layer above. The number of layers can vary according to the characteristics of the cluster, but a minimum of two layers is required. Here, we have three layers (L_0–L_1–L_2), but n possible layers can be added, in order to develop a more complex production process. Final firms (L_0agents) need one supplier from L_1, and the latter needs his own supplier from L_2, to assemble and sell the final product on the market. The market demand of the cluster's products is assumed to be fixed.

Firms differ in the quality of the goods they produce: $0.5 \leq Q < 1.0$; where Q=0.5 indicates a very bad partner for interaction inside the cluster, and Q=1.0indicates a very good one. The average quality value, Q=0.75, , is the threshold the agents use to discriminate between a good and a bad supplier.

Firms buy components from suppliers at a fixed cost, K=0.75 (thousands of euro), but the profits, U, they can make depend on supplier's quality: $U_{Li}=f(Q_{Li+1})= F * Q_{Li+1} - K$. Profits loss may be explained as if the bad quality components needs more work to be assembled and prepared for the final product.

Both L_0 and L_1 agents evaluate their suppliers, comparing the quality of the product they bought with the threshold value set at 0.75, and store these evaluations. If the product's quality exceeds that threshold, the supplier is considered good, otherwise it is labelled as a bad supplier. In an attempt to maximise their profits, firms always avoid interactions with bad suppliers, while trying to interact with the best known ones.

2.2 Information Exchange

The material exchange described above is paired in the model with an exchange of social evaluations: when the transmission of social evaluations is allowed, both leader firms and suppliers exchange information with their fellows regarding their suppliers from the level below, thus creating and taking part in a social network.

This process works only horizontally. There is no communication between agents that inhabit different levels of the cluster. Since agents exchange goods only with a specific set of suppliers, the information they acquire are only relevant inside their own level. Inside a layer, agents can play two possible roles: (1) the *Questioner* asks an Informer, i.e. another firm of the same layer, to suggest a good supplier; (2) the *Informer* provides the ID of a good supplier. Honest informers suggest their best rated supplier, whereas cheaters transmit an evaluation concerning the worse supplier, as if it was a good one.

Partner selection can then be performed in three ways, listed in their priority order:

1. *Experience-based Selection*: the best rated supplier among those already tested isc hosen
2. *Communication-based Selection*: the most frequently suggested supplier by trusted Informers is selected
3. *Random Selection*: the first available agent among the unfamiliar suppliers is selected(a sa ne scapeproc edure)

All the relevant social information are stored by the agents in three different internal repositories, which are updated and checked at run-time to take decisions regarding both suppliers and informers.

Image Table: As previously stated, firms store here the memories of their economic transactions (i.e. the actual quality value of each known supplier).

Candidates Table: Identities of potential good suppliers – without reference to his quality – suggested by a fixed percentage of Informers among the same level (10% in the current implementation of the model) are aggregated here, either in the form of a direct evaluation (*image*) or a reported evaluation, i.e. in which the source is impersonal (*reputation*). To most frequently suggested agent is selected for economic transaction, and after the transaction the information acquired are updated in the Image Table.

Informers Table: Once the information about suggested suppliers are tested and moved to the Image Table, the Informers Table is updated with the ratings of the sources of the information. If a suggested supplier is found to be bad ($Q < 0.75$), the credibility of the Informers is compromised, the agents are categorised as cheaters and further communications from them are discarded.

The presence of cheaters in the cluster set up a social dilemma. Agents acting as suppliers are able to fulfill just one economic transaction at each simulation turn.

Hence, giving away the identity of a good supplier, agents reduce the probability to interact with him in the future. False evaluations, on the other hand, have two main effects: they enhance the chances of advantageous economic transactions for the cheaters, keeping away the other firms from the good suppliers; and, at the same time, let the cheaters take advantages of the information received from truthful firms – information acquired without the costs of a possible bad economic transaction.

After a cheater is detected, cooperative firms adopt a retaliatory strategy: known cheating Questioners are provided with false evaluation even by cooperative Informers. Obviously, this behaviour depends on the type of evaluation circulating in the cluster. In the case of *reputation*, lacking an identifiable source, agents are not allowed to retaliate.

Hence, our main research question is whether the exchange of social evaluations – and what type of evaluations – can improve the economic performance of the cluster, when firms in the first two levels compete over high quality suppliers, and communication can be exploited by cheaters.

3 Results

We tested the agents' performance in terms of average quality of production, both for single layers and for the cluster as a whole. A cluster of 300 firms was so composed: 20% of agents for L_0, 40% of agents for $L_1$and 40% for L_2. Quality was assigned randomly during the set up of each simulation's run, with values distributed normally between 0.5 and 1.0. During set up was also defined the behavioural trait of the agents relevant for the communication process: honest agents were the ones who cooperated with others with truthful information about available suppliers; a fraction of the population was instead composed by Informers who spread false information, agents that we called cheaters. In each experimental setting, a different number of agents in the population was conferred the cheating trait.

We ran the experiments in three different conditions:

- *Control Condition (CC)*: no communication allowed. In this case, social information was not available and the suppliers' choice was exclusively experience-based. In this condition the behavioural trait relevant for communication was not considered: hence there were no cheaters in the population.
- *Image Condition (IC)*: agents exchanged true or false images. At the beginning of each simulation turn, agents in L_0 and L_1 were given the opportunity to collect information about available suppliers among their pairs. Contacted Informers replied according to their behavioural trait. The percentage of cheaters was set by a *cheating rate* parameter: the higher the value of the parameter the higher the number of cheaters in the cluster. Retaliation was possible: honest agents responded to cheaters, previously detected, with false information in order to punish them.
- *Reputation Condition (RC)*: agents were still allowed to communicate at the beginning of each turn, however the messages exchanged by the Informers

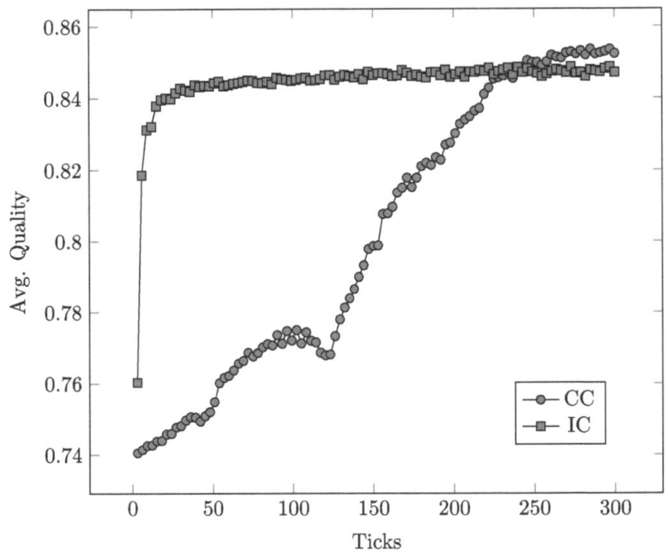

Fig. 2 Average values of quality in the Image Condition (IC) compared to the Control Condition (CC). Simulations are performed ten times in each condition with a cluster of 300 firms

contained reputational information, i.e. evaluations without an explicit source. In this case, retaliation was not allowed, since the evaluator remained unknown.

Comparing the Control Condition (CC) with the Image Condition (IC), we found that the possibility to communicate positively affects cluster's performance. In Fig. 2, agents who are not allowed to communicate randomly explore the cluster and learn the identities of the best suppliers. Since the cluster is a closed environment, i.e. there is no change over time in its composition, this learning phase comes to an end once all the suppliers are being tested. There will still be competition for the good ones, but this will not affect the economic performance of the cluster.

A different pattern arising from agents interaction is observed in the IC: when agents of the same level are allowed to communicate with one another quality values increase more rapidly, because exploring the cluster in order to obtain higher profit requires less time. However, the presence of cheaters in the communication process can alter this effect, with a profit for the cheaters, but with a great damage to the cluster as a whole (see Fig. 3).

Changing the type of evaluations exchanged among agents, cheating is still bad for the average product quality – especially for high levels of cheating rate. But in the Reputation Condition, as Fig. 4 shows, in the long run the cluster can absorb relatively high percentages of cheaters, without compromising its economic performance.

The communication algorithm proved to be robust with respect to the numbers of the agents in the cluster. What is really important is their distribution among the three levels, which in turn affects the availability of suppliers and the competition over them. When firms can chose among many suppliers (Fig. 5) we have no

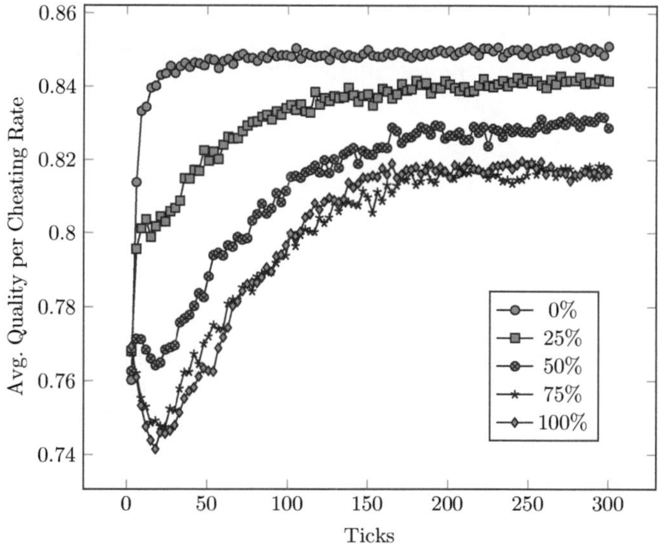

Fig. 3 Average quality per cheating rate in the two experimental conditions: Image Condition

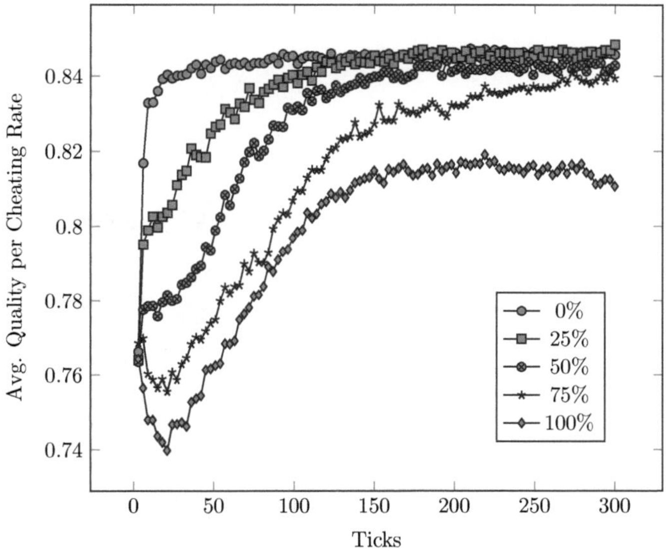

Fig. 4 Average quality per cheating rate in the two experimental conditions: Reputation Condition

difference between IC and RC. When competition is hard, however, communication in RC performs better not only in the long run, but the all cluster obtain higher profits if compared to IC (Fig. 6).

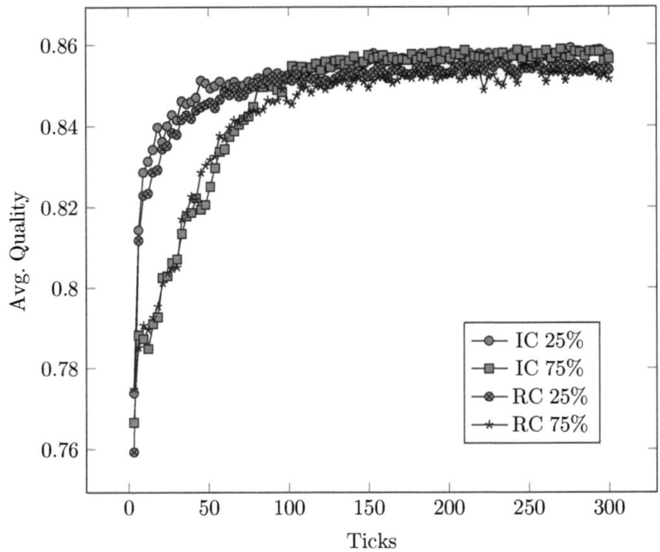

Fig. 5 Average quality of 300 firms in IC and RC for 25% and 75% cheating rate. Distribution of firms in the cluster: L_0=10%, L_1=45%, L_2=45%

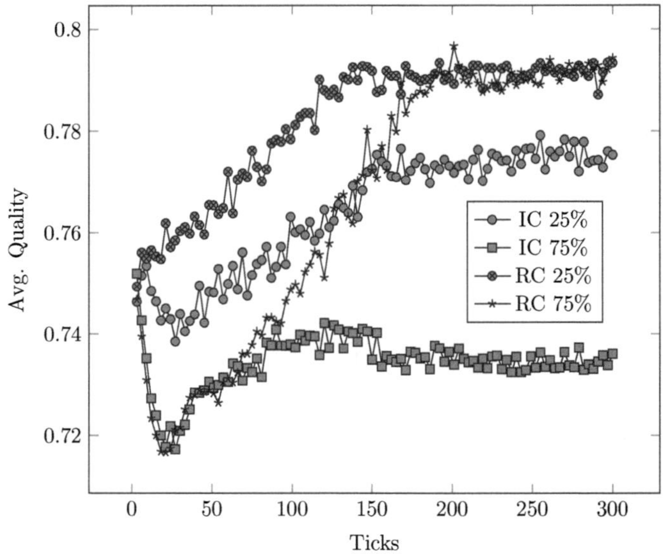

Fig. 6 Average quality of 300 firms in IC and RC for 25% and 75% cheating rate. Distribution of firms in the cluster: L_0=30%, L_1=35%, L_2=35%

4 Concluding Remarks

The present study sought to test the effects of two different kinds of social evaluations in an artificial cluster, adding to previous studies that applied the social and cognitive theory of reputation to other settings, both natural and artificial [6]. We suggested that image and reputation, although closely related, are distinct objects, with different aims, functions and effects. We used the "small-world" of industrial clusters as a test bed of our theory, given the importance that reputational concerns have in this context. Material exchanges are usually supported and even improved by the social network of individuals and firms acting into a cluster: the merging of economic structure and social community makes the exchange of social evaluations especially relevant to isolate cheaters, prevent frauds between cluster's actors and preserve quality of the single firms and of the entire cluster.

In order to test our predictions about the positive effects of communication on firms' economic performances, we designed an artificial cluster with companies grouped into three layers that trade products and exchange social evaluations. In this artificial cluster we tried to figure out how social information may affect the search for good partners and whether image and reputation make a difference to economic performances of both single firms and district as a whole. Our results showed that communication matters: compared to control condition, communication positively affected cluster's performance. Firms receiving reliable information about potential partners easily found good suppliers, compared to firms that were systematically cheated by their fellows. Furthermore, modelling reputation as rumors – i.e. evaluations where the source is unknown – we were able to preserve the benefits of communications for low level of cheating rate. In other words, reputation prevented retaliation, thus avoiding generalized punishment that would have lowered firms' profits.

We acknowledge that further improvements are needed, regarding both agents' refinement and cluster's structure. Although very elementary, our model allows to verify theoretic predictions about the different effects of image and reputation and to relate them with the economic performance of an idealized industrial district. Future directions of work will include introduction of communication flows between levels, refinement of firms' economic structure and testing for other social control mechanisms, as for instance ostracism. On the one hand, the lack of a true cognitive architecture prevented the possibility of exploring more interesting ways of implementing agents' decision making, since agents' strategies varied only according to the cheating rate. The hyper-simplified economic structure, however, allowed us to analyze what happens at the macro-level, linking it directly to agents' actions.

Acknowledgements This work was partially supported by the Italian Ministry of University and Scientific Research under the Firb programme (Socrate project, contract number RBNE03Y338), and by the European Community under the FP6 programme (eRep project, contract number CIT5-028575), and by the European Science Foundation under the EUROCORES Programme TECT: The Evolution of Cooperation and Trading (SOCCOP project).

References

1. Albino V, Carbonara N, Giannoccaro I (2003) Coordination mechanisms based on cooperation and competition within industrial districts: an agent-based computational approach. J Artif Soc Soc Simulat 6(4). http://jasss.soc.surrey.ac.uk/6/4/3.html
2. Alexander RD (1987) The biology of moral systems. Aldine de Gruyter, New York
3. Becattini G (1990) The Marshallian industrial district as socio-economic notion. In: Pyke F, Becattini G, Sengenberger W (eds) Industrial districts and inter-firm cooperation in Italy. International Institute of Labour Studies, Geneva, pp 37–51
4. Brenner T (2001) Simulating the evolution of localised industrial clusters – an identification of the basic mechanisms. J Artif Soc Soc Simulat 4(3). http://jasss.soc.surrey.ac.uk/4/3/4.html
5. Castelfranchi C, Conte R, Paolucci M (1998) Normative reputation and the costs of compliance. J Artif Soc Soc Simulat 1(3). http://jasss.soc.surrey.ac.uk/1/3/3.html
6. Conte R, Paolucci M (2002) Reputation in artificial societies: social beliefs for social order. Springer, Heidelberg
7. Dellarocas C (2003) The digitization of word of mouth: promise and challenges of online feedback mechanisms. Manag Sci 49(10):1407–1424
8. Fehr E, Gachter S (2000) Fairness and retaliation: the economics of reciprocity. J Econ Perspect 14(3):159–181
9. Gintis H, Smith EA, Bowles S (2001) Costly signaling and cooperation. J Theor Biol 213(1):103–119
10. Gluckman M (1963) Papers in honor of Melville J. Herskovits: gossip and scandal. Curr Anthropol 4(3):307–316
11. Granovetter M (1985) Economic action and social structure: the problem of embeddedness. Am J Sociol 91(3):481–510
12. Henrich J (2006) Cooperation, punishment, and the evolution of human institutions. Science 312(5770):60–61
13. Henrich J, Boyd R (2001) Why people punish defectors. Weak conformist transmission can stabilize costly enforcement of norms in cooperative dilemmas. J Theor Biol 208(1):79–89
14. Karlsson C, Johansson B, Stough R (2005) Industrial clusters and inter-firm networks (new horizons in regional science series). Edward Elgar Publishing, Cheltenham
15. Lane D (2002) Complexity and local interactions: toward a theory of industrial districts. In: Curzio QA, Fortis M (eds) Complexity and industrial clusters: dynamics and models in theory and practice. Physica-Verlag, Heidelberg, pp 65–82
16. Nowak MA, Sigmund K (1998) Evolution of indirect reciprocity by image scoring. Nature 393(6685):573–577
17. Panchanathan K, Boyd R (2004) Indirect reciprocity can stabilize cooperation without the second-order free rider problem. Nature 432(7016):499–502
18. Sabater J, Paolucci M, Conte R (2006) Repage: reputation and image among limited autonomous partners. J Artif Soc Soc Simulat 9(2). http://jasss.soc.surrey.ac.uk/9/2/3.html
19. Sommerfeld RD, Krambeck HJ, Semmann D, Milinski M (2007) Gossip as an alternative for direct observation in games of indirect reciprocity. Proc Natl Acad Sci USA 104(44): 17,435–17,440
20. Squazzoni F, Boero R (2002) Economic performance, inter-firm relations and local institutional engineering in a computational prototype of industrial districts. J Artif Soc Soc Simulat 5(1). http://jasss.soc.surrey.ac.uk/5/1/1.html
21. Wedekind C, Milinski M (2000) Cooperation through image scoring in humans. Science 288(5467):850–852
22. Wilson DS, Wilczynski C, Wells A, Weiser L (2000) Gossip and other aspects of language as group-level adaptations. In: Heyes C, Huber L (eds) The evolution of cognition. MIT Press, Cambridge

Part III
Modeling Approaches and Programming Environments

Injecting Data into Agent-Based Simulation

Samer Hassan, Juan Pavón, Luis Antunes, and Nigel Gilbert

Abstract Many agent-based models, use standard distributions in several steps of the design: configuring the initial conditions of simulations, distributing objects spatially, and determining exogenous factors or aspects of the agents' behaviour. An alternative approach that is growing in popularity is data-driven agent-based simulation. This paper encourages modellers to continue this trend, discussing some guidelines for finding suitable data and feeding models with it. In addition it proposes to merge the principles of microsimulation into the classical logic of agent-based simulation, adapting it to the data-driven approach. A case study comparing the two approaches is provided.

Keywords Agent-based modelling • Data-driven • Microsimulation • Quantitative data • Random initialisation • Social simulation

1 Introduction

Many Agent-Based Models (ABM) aim to simulate some real-world phenomenon and their validation is usually driven by empirical data. For the purpose of this paper, we assume that a 'good' model of some target phenomenon is one that shows

---

S.H assan(⊠)
GRASIA, Universidad Complutense de Madrid, Madrid, Spain
CRESS, University of Surrey, Surrey, UK
e-mail: samer@fdi.ucm.es

J.P avón
GRASIA,U niversidadC omplutensed eM adrid, Madrid,Sp ain
e-mail:j pavon@fdi.ucm.es

L.A ntunes
GUESS/LabMAg/Universidaded eL isboa, Lisboa,Por tugal
e-mail: xarax@di.fc.ul.pt

N. Gilbert
CRESS, University of Surrey, UK
e-mail:n.gilbe rt@surrey.ac.uk

K. Takadama et al. (eds.), *Simulating Interacting Agents and Social Phenomena: The Second World Congress*, Agent-Based Social Systems 7,
DOI 10.1007/978-4-431-99781-8_13, © Springer 2010

the same behaviour as the target. However, the initial conditions of a model usually do not attempt to reproduce the real world. Most often, the simulation begins with values taken from a uniform random distribution. But there are many cases where the choice of initial conditions can affect the output of the model and where a uniform random distribution is a poor choice.

There are some well known examples of ABMs where the modelling has been closely linked to empirical data [2]. One is the model of the extinction of the Anasazi civilisation, in which empirical data are used to improve the fit between the simulation and the observed history. In this example, the exogenous factors (environmental variables) are not randomized, although the initial conditions are [3]. Another example is the water demand models of [4, 5], in which data about household location and composition, consumption habits and water management policies are used to steer the model, with good effects when the model is validated against actual water usage patterns. A third case is Hedström's model of youth unemployment [14] in which data from surveys are imported and regression equations are used to calculate transition probabilities. Another interesting model is [7] because it used qualitative data from interviews. From a broader point of view, there are examples such as pedestrian flow modelling using spatial data [1] and simulations of markets such as that of the electricity market [15]. An example of how an ABM can be improved by introducing data, in contrast with the random approach to initialisation, is a study of the Eurovision song contest [6]. This considers voting in a popular music contest in Europe, and begins with the hypothesis, "over a sufficiently long period of time the results of the Eurovision contest would approximate to random". If the hypothesis were true, a simulation with random initial conditions and random voting schema should approach the real situation. But actually it does not. It is shown that introducing empirical data, such as the distance between countries (if a country is closer, people are more likely to vote for it) or a measure of the similarity of their cultures, improves the results of the simulations.

These examples show how initialisation can be addressed by gathering data and feeding the model with it. There are some similarities with a technique called microsimulation (also known as microanalytic simulation) [10]. Microsimulation focuses on the simulation of the behaviour of individuals over time. The individuals are initialised with empirical data (usually derived from a sample survey). The simulation consists of repeatedly changing the simulated individuals according to a set of transition probabilities and transition rules (ideally, both extracted from empirical data). However, microsimulation does not model interactions between individuals, each of whom is considered in isolation.

This paper encourages ABM designers to continue the data-driven trend, by merging some concepts taken from microsimulation into ABM. First, the classical logic of simulation and some of the problems arising from abstract models are reviewed. In Sect. 3 possible sources of data for use by modellers are discussed. In Sect. 4 the alternative approach is outlined, while in Sect. 5 its main difficulties are examined. The approach can contribute to obtaining simulation results that are closer to observations of the corresponding target, as will be shown in a case study described in Sect. 6. The final section concludes with a few tentative guidelines.

2 The Classical View

2.1 *The Logic of Simulation*

Agent-based modelling is founded on a methodology that has been described as a "logic of simulation" [9]. This logic, shown in diagrammatic form in Fig. 1, is a representation of the classical scientific experiment applied to the simulation. The Target is the observed phenomenon. As a result of a process of Abstraction, a Model, a simplification of this phenomenon, can be obtained. This Model, in this case an Agent-Based Model, can be simulated to obtain results, the Simulation data. A process of Data gathering (qualitative, quantitative, or both) can be used to extract the Collected data from the Target. The comparison of this data and the simulation output allows a process of validation. If there is structural similarity between them, the ABM is considered a good representation of the phenomenon. If there is not, the model should be modified and the simulation repeated until the output fits the gathered data.

2.2 *Issues with Abstract ABMs*

In this classical approach, modellers seek generality through a high level of abstraction. Thus, instead of empirical data which is specific in space and time, they tend to

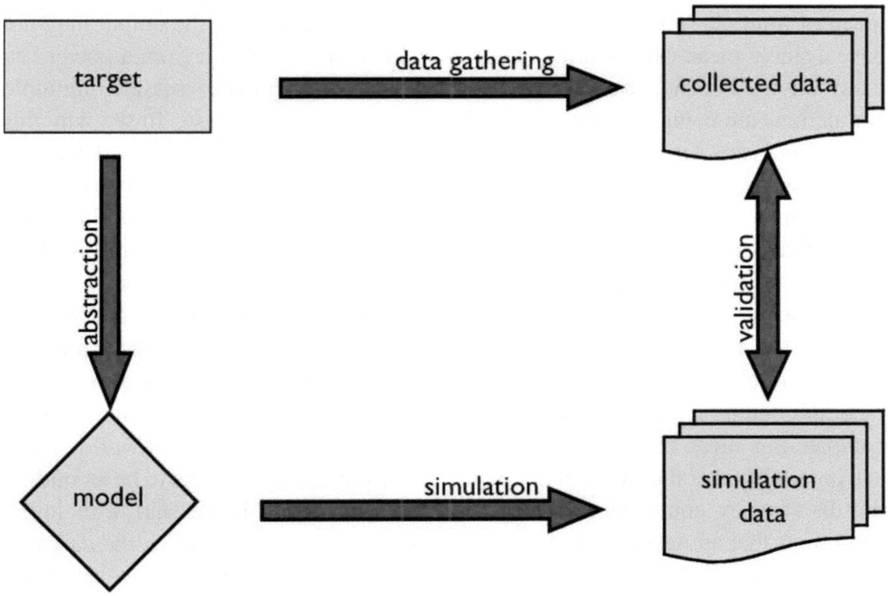

Fig. 1 Diagram showing the classical logic of simulation (after [9])

use standard distributions in several steps of the design: configuring the initial conditions of simulations, distributing objects spatially, and determining exogenous factors or aspects of the agents' behaviour. The advantage of using an abstract model and random values is that the model can be considered to be more general, applying not just to one specific case, but to any circumstances within the bounds of the stochastic distributions used to obtain parameter values.

The most popular distribution is the uniform random distribution, which is commonly applied to generate a model's initial conditions. The typical procedure is to run a series of simulations (each with a different starting random seed value) and aggregate their outputs into a mean. This is an appropriate method to check the relationships among a set of parameters in a model. However, it does not ensure that the output cannot be improved with other initial conditions, especially when there is a need to compare with real systems and precise data.

For some parameters it is more appropriate to use distributions other than the uniform random. For instance, a Gaussian distribution usually fits empirical data on individual income quite well. But to ensure that it is the right distribution for the target situation, we have to have recur to real data and in that case, why not use those data directly, rather than abstracting them into an ideal-typical distribution?

When the correct statistical distribution is not known, it is better to use one or more empirical distributions. Or, a hypothetical but typical set of data could be used. The problem is that 'typical' is hard to define formally. Statistical methods aim to define that notion. Another fundamental problem with probability distributions is that while they are good at describing aggregate behaviours, especially from an a posteriori perspective, they do not provide the *reasons* that may cause individual behaviour.

Further problems arise from the implications of the procedure of comparing the mean of multiple runs with one observation of the target. First, the output may not have a stable mean (this is often the case when values are drawn from a power law distribution). Second, even when the mean is the appropriate measure, multiple simulations are being compared with observations of a single case. To see why this can be an issue, let us assume that at least some observable elements of the real world are stochastic. Then the one instance of the real world that actually exists can be thought of as a random selection from a population of possible worlds. That means that, while the most probable case is that the real world has the same attribute values as the means of the values in all possible worlds, it is also quite likely that the real world value is not close to the mean and certainly possible that it is an outlier, far from the mean. Now suppose that due to some happy chance, we have a model that accurately represents the real social processes. We initialise the model with random conditions, run it many times and calculate the mean behaviour. We then compare this mean behaviour with the behaviour observed in the real world. There is a possibility that the two will not match. If the real world happens to be an outlier, the discrepancy could be very large. On the other hand, if we start with initial conditions that are taken from data, even if the real world is an outlier, the data will to some degree move the model in the direction of the real world, and we are much more likely to find a match between the model and the observed data.

Since we advocate the use of data not only for validation, but in other phases of the simulation development, we must pay close attention to ensuring that they are representative of the universe for which we are designing the model. In the following sections we provide some procedures for how to handle data for the purposes of social simulation.

3 Sources of Data

Once it has been decided that data will be used to drive the simulation, the next questions are, what type of data, and where could the data be obtained?

It is desirable to have data from some representative sample of the target population. In practice, this usually means survey data from a large random sample of individuals, although it needs to be recognised that large representative samples, while statistically advantageous, also have some disadvantages:

1. If the sample is large, it is likely that the researcher will not be the person who designs or carries out the survey. More likely, the data will come from a government or market research source. This means that the survey will probably not include exactly the right questions phrased in the right way for the researcher's interests, and compromises will have to be made.
2. If the sample is random, it is unlikely that it will include much or any data about interconnections and interactions between sample members, so studying networks of any kind is likely to be impossible. This can be a serious problem when the topic for investigation concerns matters such as the diffusion of innovation or information, or friendship relations.
3. Some data are inherently qualitative and not easily gathered by means of social surveys. For example if one is interested in workplace socialisation (e.g. [18]), a survey of employees is a very crude and ineffective method as compared with focused interviews, focus groups or participant observation (for more details on theses tandardm ethodsofs ocialre search,s ee[8]).

Despite these disadvantages, survey data can be valuable. It is particularly valuable when it is collected from panels, i.e., if the same individuals are interviewed several times at intervals, such as every year. Panel studies are more or less the only way of collecting reliable data about change at the individual level. Panel data can be used to calculate transition matrices, that is, the probability that an individual in one state changes to another state (e.g. the probability of unemployment). With a sufficient amount of data, one can calculate such transition matrices for many different types of individual (i.e., for many different combinations of attributes). So for example, it becomes possible to calculate the rates of unemployment for young men, old men, young women and old women. However, if one tries to take this too far – differentiating according to too many attributes – the reliability of the computed probabilities becomes too low, because there will be too few cases for each combination of attributes. These probabilities provide the raw material for constructing probability distributions that may be used to simulate the effect of the passage of time on individuals.

We have stressed the importance of obtaining data repeatedly over periods of time. This is because generally agent-based models are concerned with dynamical processes, and snapshots of the situation at one moment in time are of limited value and can sometimes even be misleading as the data basis for such models. While panel survey data is relatively rare compared with cross-sectional data, other forms of data collection about social phenomena are often more attuned to measuring processes. This is particularly the case with ethnography where the researcher observes a social setting or group continuously over periods of days, weeks or months. A third form of data collection is to use official documents, internet records and other forms of unobtrusive data that are generated by participants as a byproduct of their normal activities, but that can later be gathered by researchers. Examples are newspaper articles, web pages, and government reports. In these cases, it is often possible to collect a time series of data (e.g. using the Internet Archive http://www.archive.org/ to recover the history of changes to a web site) and thus to examine processes of change.

Regardless of whether the data is quantitative or qualitative, it is often the case that they do not have to be collected afresh, but rather that data previously collected by another organisation, possibly for another purpose, can be used. Enormous quantities of survey and administrative data are stored in national Data Archives (European archives are listed at http://www.nsd.uib.no/cessda/archives.html) and increasingly Archives are extending their scope to include qualitative data (e.g. in-depth interviews) as well (see for example, http://www.esds.ac.uk/qualidata/).

4 The Data-Driven Flow: Adapting the Logic

The aim of this section is to propose an alternative to the classical logic described in Sect. 2.1. The design is an idealisation of what will normally be a less clear cut process. It could be specially useful in contexts where there are quantitative or qualitative empirical data from existing sources, or at least the possibility of collecting samples of such data. It uses ideas that originated with microsimulation.

Microsimulation [10] has traditionally been used in areas where it is easy to obtain quantitative data, in the form of surveys and censuses (for the initialisation of individual units) and equations or rules (for defining agent behaviour).

Although microsimulation has been successful in some problem domains such as traffic modelling and econometrics, it has been difficult to apply in social domains that are not so well structured or where there are important dependencies between agents. Microsimulation is unable to model interactions between agents, an area where agent-based modelling is pre-eminent. Nevertheless, some aspects of microsimulation, such as basing the simulation on representative survey samples and using probability transition matrices to determine changes in the values of

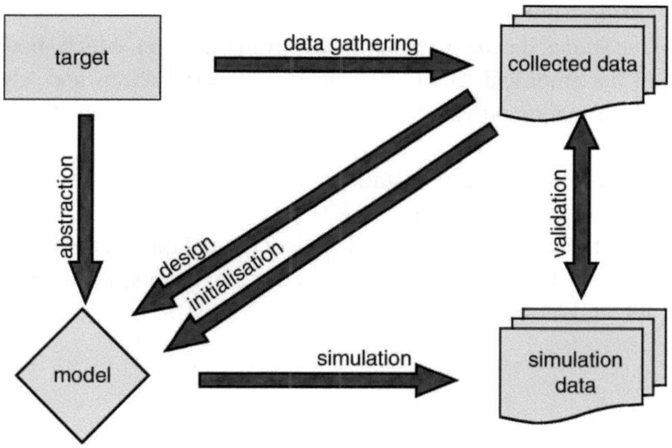

Fig. 2 Diagram showing a modified logic of simulation for data-driven modelling

agent parameters, can usefully be applied to the design of ABM. Agent-based models usually follow an event-based rules approach rather than using transition probabilities. However, the limitations of modelling or the lack of sufficient data frequently make it difficult to implement explicit rules and therefore they have to turn to other solutions, one of which is to use transition probabilities, which represent implicit rules. Qualitative information, although rarely used in ABM, can also be introduced [18].

Adopting this approach it is possible to reformulate the classical stages of the logic of simulation, importing elements from microsimulation and adapting it to better suit the data-driven modelling approach. The main change is the focus on collected data. In the classical logic of simulation presented in Sect. 2.1, the data gathering could be done after building the model and the simulation, because the data were used just for validation.

Figure 2 reproduces Fig. 1, but with two new arrows, representing the use of collected data to design and initialise the model. The new flows force the data gathering to be done before the simulation. Building the model is not finished until the abstraction, data-driven design and initialisation are all completed. Only then can the simulation be executed and the output obtained. The last stage, the validation process, must be done with data not used previously in initialisation.

Although these stages are presented in a linear way, the design and development process is usually carried out in an iterative manner. Thus, there may be a need for feedback from the results of the validation stage, forcing changes to the design of the ABM.

Therefore, the diagram can be condensed into these sequential steps (which match the six arrows in Fig. 1):

1. *Data gathering* from the social world
2. *Design of the model*, the *abstraction* process from the target, which should be guided by some of the empirical data (e.g. equations, generalisations and 'stylised facts', qualitative information provided by experts) and by the theory and hypotheses of the model
3. *Initialisation* of the model with static data (e.g. surveys and census)
4. *Simulation* and output of results
5. *Validation*, comparing the output with the collected data. The data used in validation should not be the same as that used in earlier steps, to ensure the independence of the validation tests from the model design

5 Discussion and Difficulties of the Approach

The application of this procedure can present some difficulties. For some models, especially those at a high level of abstraction, appropriate data may be impossible to obtain. Another problem, common also to microsimulation, is the requirement for large volumes of detailed data about individuals. Although a data gathering effort is frequently required for validation, it may not have the special intensity required here. This additional costs may not be worthwhile in certain cases, in spite of the expected improvement of results.

If the data are extracted from several sources, it can be quite difficult to merge it to avoid inconsistencies. And handling huge amounts of data makes still more complicated the process of deciding what is relevant. In all cases, representativeness and relevance are important criteria in selecting data manipulation procedures.

Sometimes, the lack of data stems merely from the absence of suitable surveys and other data sources. Sometimes, the problem is more fundamental. For example, agent characteristics such as their emotional states are unobservable. In some models, the agents' current state depends on their previous circumstances (this is the case, for example, in models which incorporate path dependencies, or where agents have memory). However, it is rare for such histories to be recorded systematically in representative surveys. It is also often hard to obtain information regarding networks and micro-interaction processes, unless one is dealing with very particular domains such as virtual communities where data are recorded as a side effect of electronic interactions [17].

Some of these problems can be overcome or worked around. For example, if we want to simulate a married couple, we can find a wife in a survey based on a random sample of individuals, but we also need an agent to represent her husband. Since the data are taken from a random sample, it is unlikely that the husband will also be in the survey. Strategies for dealing with this include creating an artificial 'husband', not based on anyone in the sample; or 'marrying' the woman to a different, married man in the sample.

6 A Case Study: the Mentat Model

6.1 Context of the Model

The aim of the Mentat agent-based model [13] is to understand the evolution of social and moral values in Spain from 1980 to 2000. This period is interesting because of the substantial shift in moral values corresponding to the transition from a dictatorship to a consolidated democracy. The almost 40 years of dictatorship finished in 1975, when the country was far from the European average on most indicators of progress, including the predominant moral values and modernisation level. The observed evolution of moral values since then is analogous to that found in its EU partners, but the changes in Spain developed with a special speed and intensity. The main factor proposed to explain the observed changes is demographic: the change in the age structure of the population and the influence of a younger generation. The Mentat model aims to simulate the effect of cross-generational changes, focussing on these "vertical" rather than on "horizontal" influences.

The Mentat model hypothesises that values are influenced by a range of factors, including demography, economy, political ideology, religiosity, family and friend relationships, reproduction patterns, and stage in the life course. We shall use the model to examine the effect of initialising it with empirical data, as compared with a version initialised using a random distribution. The behavioural rules at the individual level are the same in both versions. To simplify the comparison still further, we reduce the number of objective variables to the one most critical: age. Its distribution will determine the demography of the system: agents die when they are old, they search for a partner in youth, they have more or less chance to have a child depending on age, etc. Both versions of the ABM will then be validated against additional empirical data (not previously used in model initialisation).

The simulation has been configured with a population of 3,000 agents and simulated for a period of 20 years (from 1980 to 2000). The agents are able to communicate, make friends, establish couple relationships, and reproduce. They form a network where the nodes are the individuals and the links can be of type 'friend' or 'family' (couple, parents, children). The more friendships exist, the more couples and families will be formed (as the partner is chosen from friends). The model includes age-related probabilities of having children (for example, a woman in her forties will have less chance than a 23 years old); regression equations to determine whether an agent searches for a partner or not; and time-varying transition matrices for life expectancy and the fertility rate (the birth rate in Spain fell from 2.2 in 1980 to 1.19 in 2000).

6.2 The Randomly Initialised Version: Mentat-RND

The version of the ABM with random initial conditions has been named Mentat-RND, while the one with empirically based initialisation is Mentat-DAT. Both have exactly the same structure except for the source of the ages of the initial agent population. In Mentat-RND this attribute has been assigned using a uniform random distribution in the range [0, 75].

The output of the system consists of several indicators directly affected by the demographic model and the population pyramid (age distribution). We monitor the percentage of old people, the ratio of single to married agents, and the overall population growth rate (determined by the change in the number of couples and their age).

The system's output is unstable, with noticeable changes between executions, so an aggregation measure is needed. The model was executed 15 times and the mean of each indicator calculated. The results are compared with empirical data and with the results of Mentat-DAT.

6.3 The Version Initialised with Data: Mentat-DAT

The agents in Mentat-DAT are initialised using data from the Spanish census, research studies and sample surveys [16]. The basic input is from the Spanish sample of the 1980 European Values Survey (EVS). The data provide a range of variables, including demographics, attitudes and financial information, for a representative sample of 2,303 individuals surveyed in 1980. The data are used to generate a simulated population with the same statistical distributions of the main parameters as the whole Spanish population. Consequently, the population pyramid in the model is similar to the real one in Spain in the 1980s.

While Mentat-DAT is initialised using data from the 1980 EVS, the outputs from it (and Mentat-RND) after 10 and 20 simulated years are compared with data drawn from the 1990 and 1999/2000 European Values Surveys. The three sweeps of the EVS thus provide independent data sets for initialisation and for validation.

6.4 Comparison of Outputs

In this section we compare the results from Mentat-RND (random initialisation) and Mentat-DAT (data initialised version), contrasting them with data from the Spanish Population Census and the 1990 and 1999/2000 EVS. The values of the selected output indicators for the two versions of the model are shown in Table 1. A deeper analysis of the evolution of the main parameters can be found in [11]. The values for Mentat-RND are averaged over 15 executions to allow for stochastic variations in its output. Mentat-DAT is almost stable between executions because of its fixed initialisation and so the means shown are based on only five runs.

Consider first the proportion of older people. The Census shows that this has been growing, starting at 18% in 1980 and reaching 21% by 1999. Mentat-RND

Table 1 Validation: comparison between EVS, the random initialised version and the data-driven version

	EVS/Census*			Mentat-RND			Mentat-DAT		
	1980	1990	1999	1980	1990	1999	1980	1990	1999
% 65+ years	16*	18*	21*	19	24	29	15	19	24
% Single	28	29	29	–	45	37	–	42	35
% Population Growth	–	–	+8%*	–	–	+10.1%	–	–	+7.2%

*Source: Spanish Population Census for the years 1981, 1991 and 2001

begins with almost the correct figure in the (simulated) year 1980, but the rate of growth is much faster than it should be. On the other hand, Mentat-DAT shows a closer fit to the empirical data.

The observed proportion of single people is steady over time. The number of couples in the ABM is directly proportional to the number of friendship links, so the ratio of single to married agents is a good measure of the cohesion of the network. In Mentat-DAT, the attributes of the individual agents are initialised from the 1980 EVS data, but not the couples, as there is no information about links between members of the sample in the EVS. The simulation must therefore start by creating such links to build the network structure. Only after some execution steps does the proportion of couples converge to a steady state. We can see that Mentat-DAT is again closer to the survey data than the randomly initialised version. Continuing both simulations beyond 20 years allows us to observe a convergence to a proportion of single people in the range [28, 30], but this is reached more slowly by Mentat-RND.

For the case of population growth, the randomized version generates a rate of 10.1%, higher than the Census (8%), while the data-driven version has a growth rate slightly lower (7.2%) than the Census. Overall, the data-driven Mentat-DAT provides a closer fit to the empirical data than the randomly initialised Mentat-RND for all three of these parameters.

7 Concluding Remarks

The motivation for this paper was a concern about the use of random initialisation in ABM, and the possibility of basing models more closely on empirical data. The approach described here merges some aspects of Microsimulation with ABM. The Mentat model [11] was used to illustrate that feeding a model with empirical data can improve the fit between it and the observed social world: for example, its internal dynamics, its macro-level behaviour, and the structure of the networks linking agents.

In this paper, we have discussed some of the issues that need consideration when injecting data into an ABM. We have suggested that exposing a model to data does not have to be left to the final, validation step, but has value at the very beginning of the modelling process. Thus, some changes were introduced to the methodological steps of the classical logic of simulation [12]. As a result of the experience with the Mentat model, the following guidelines can be suggested:

- It is valuable to explore the problem background, focusing not only on the theoretical literature, but also on the availability of data
- It is worthwhile to compare different collections of data and conclusions from diverse sources to give a stronger foundation to the model
- The most valuable data are those that provide repeated measurements, preferably taken from the same respondents (as in a panel survey)
- The ABM should be designed so that it generates output that can be compared directly with empirical data
- If the data are available, it is recommended to simulate the past and validate with the present, as was done in the case study

The effect of applying these suggestions would be to connect the majority of agent-based models more closely to the social world that they intend to simulate, at the cost of the extra effort and complication involved in injecting data into the simulation.

Acknowledgments We acknowledge support from the projects: *NEMO: Network Models, Governance and R&D collaboration networks* funded by the European Commission Sixth Framework Programme – Information Society and Technologies – Citizens and Governance in the Knowledge Based Society, and *Agent-based Modelling and Simulation of Complex Social Systems (SiCoSSys)*, supported by Spanish Council for Science and Innovation, with grant TIN2008-06464-C03-01. We would also like to thank the anonymous reviewers for their valuable comments.

References

1. Batty M (2001) Agent-based pedestrian modeling. Environ Plann B Plann Des 28:321–326
2. Boero R, Squazzoni F (2005) Does empirical embeddedness matter? Methodological issues on agent-based models for analytical social science. J Artif Soc Soc Simulat 8(4)6. http://jasss.soc.surrey.ac.uk/8/4/6.html
3. Dean JS, Gumerman GJ, Epstein JM, Axtell RL, Swedlund AC, Parker MT, McCarroll S (2000) Understanding Anasazi culture change through agent-based modeling. Oxford University Press, Oxford, pp 179–205
4. Edmonds B, Moss S (2005) From KISS to KIDS – an 'anti-simplistic' modeling approach. http://hdl.handle.net/2173/13039
5. Galán JM, López-Paredes A, Olmo R (2010) An agent based model for domestic water management in Valladolid metropolitan area. Water Resour Res 45(5):w05401
6. Gatherer D (2006) Comparison of eurovision song contest simulation with actual results reveals shifting patterns of collusive voting alliances. J Artif Soc Soc Simulat 9(2)1. http://jasss.soc.surrey.ac.uk/9/2/1.html
7. Geller A (2008) Power, resources and violence in contemporary conflict: artificial evidence. In: Second world congress of social simulation, Washington, DC
8. Gilbert N (2008) Researching social life, 3rd edn. SAGE Ltd., London
9. Gilbert N, Troitzsch KG (1999) Simulation for the social scientist, 1st edn. Open University Press, UK

10. Gupta A, Kapur V (2000) Microsimulation in government policy and forecasting. North Holland, Amsterdam
11. Hassan S, Antunes L, Arroyo M (2008) Deepening the demographic mechanisms in a data-driven social simulation of moral values evolution. In: MABS 2008: multi-agent-based simulation. LNAI: Lecture notes in artificial intelligence. Springer, Lisbon
12. Hassan S, Antunes L, Pavón J, Gilbert N (2008) Stepping on earth: a roadmap for data-driven agent-based modelling. In Proceedings of the fifth conference of the European social simulation association (ESSA08), Brescia, Italy
13. Hassan S, Pavón J, Arroyo M, Leon C (2007) Agent based simulation framework for quantitative and qualitative social research: statistics and natural language generation. In: Amblard F (ed) ESSA'07: fourth conference of the European social simulation association, Toulouse, France, pp 697–707
14. Hedström P (2005) Dissecting the social: on the principles of analytical sociology. Cambridge University Press, Cambridge
15. Nicolaisen J, Petrov V, Tesfatsion L (2001) Market power and efficiency in a computational electricity market with discriminatory double-auction pricing. IEEE Trans Evol Comput 5:504–523
16. Pavón J, Arroyo M, Hassan S, Sansores C (2008) Agent-based modelling and simulation for the analysis of social patterns. Pattern Recogn Lett 29:1039–1048
17. Taraborelli D, Roth C, Gilbert N (2008) Measuring wiki viability (ii). Towards a standard framework for tracking content-based online communities (white paper)
18. Yang L, Gilbert N (2007) Getting away from numbers: using qualitative observation for agent-based modelling. In: Amblard F (ed) ESSA'07: fourth conference of the European social simulation association, Toulouse, France, pp 205–214

The MASON HouseholdsWorld Model of Pastoral Nomad Societies

Claudio Cioffi-Revilla, J. Daniel Rogers, and Maciek Latek

Abstract Computational modeling of pastoralist societies that range as nomads over diverse environmental zones poses interesting challenges beyond those posed by sedentary societies. We present HouseholdsWorld, a new agent-based model of agro-pastoralists in a natural habitat that includes deserts, grasslands, and mountains. This is the paper-of-record for the HouseholdsWorld model as part of a broader interdisciplinary project on computational modeling of long-term human adaptations in Inner Asia. The model is used for conducting experiments on socio-environmental interactions, social dynamics experiments, and for developing additional models with higher levels of social complexity.

Keywords Social simulation • Agent-based modeling • Computational social science • Pastoral • Nomads • Inner Asia • Mongolia • MASON toolkit • Coupled socio-natural systems • Climate change • Social complexity

1 Introduction: Motivation and Background

From a comparative social perspective, nomadic societies – Mongols, Huns, Roma, Beduins, among others – exhibit a defining or characteristic form of collective spatial mobility that distinguishes them from the much larger class of sedentary

C.C ioffi-Revilla(✉) and M. Latek
Department of Computational Social Science, Center for Social Complexity MSN 6B2, Krasnow Institute for Advanced Study, George Mason University, 4400 University Drive, Fairfax, VA 22030,U SA
e-mail:c cioffi@gmu.edu;m latek@gmu.edu

J.D.R ogers
Department of Anthropology, National Museum of Natural History, NHB 112, 10th and ConstitutionA venue,W ashington,D C 22013,U SA
e-mail:roge rsd@si.e du

K. Takadama et al. (eds.), *Simulating Interacting Agents and Social Phenomena: The Second World Congress*, Agent-Based Social Systems 7,
DOI 10.1007/978-4-431-99781-8_14,

societies. While the drivers of nomadism can differ across societies, as well as through time, a common cause of spatial displacement arguably involves the need to follow herds of animals or other valued resources. In turn, and crucially, herds follow the changing environment according to a yearly cycle and other longer-term weather or climate patterns. Annual seasons determine local migrations; decadal or longer climate cycles can also determine inter-regional or other long-distance migrations. The number of natural habitats or ecosystems traversed by a nomadic group may be taken as a measure of its mobility, in addition to physical distance.

In Inner Asia and the Eurasian steppe, pastoral nomadic societies have played an influential role in shaping world history, at least since the rise of the Xiong-nu (Hünnü) during the late third century and early second century BCE, in the Ordos region and far beyond.[1] As described in the Chinese classic *Shih chi* (*Shiji*) by the Han court scribe Ssu-ma Ch'ien:

Each group had its own area, within which it moves about from place to place looking for water and pasture ([26]: 163–164).

Interestingly, nomadism also has effects on political organization and forms of governance, given the challenges and opportunities posed by collective mobility. As a result, nomadic societies have evolved their own adapted versions of chiefdoms, states, and empires [22]. The spatial mobility of institutions is a common necessity in nomadic societies, whereas it is rare for sedentary societies (e.g., moving a capital city or administrative center). Nomadic households are the fundamental and simplest building blocks at the bottom of the hierarchy of human groups that constitute a polity and its governance institutions. In turn, households belong to clans and tribes, which – in Inner Asia and other world regions populated by pastoral nomadic societies – in turn form confederations and other higher-level polities. Membership in such supra-household social groupings (e.g., clans, tribes, sects, and other communities) is key to understanding social organization and dynamics, including patterns of war and peace.

Social simulations of nomadic and pastoralist societies provide unique computational models and virtual laboratories for testing generative theories [11] of social complexity among interactive and mobile agents, as well as for exploring and discovering new patterns of socio-environmental interactions [16]. In addition, the possibility of conducting *in silico* experiments is especially valuable, because household interaction and mobility often raise a number of "what if" questions associated with contingencies of time and space (e.g., the rise of the Mongol empire in the thirteenth century, not earlier in history or elsewhere in Asia). Theoretical development, discovery, and experimentation provide the basic motivation for the model described in this paper.

[1]The influence of pastoral nomadic societies from the steppes of Eurasia on world history antedates the Xiong-nu period and the opening of the Silk Roads, as far back as 1000 BCE and the Scythians: "The Scythians and the related Sarmatians are the first steppe nomads of whom we have any real knowledge, although the Romans had long contact with the Parthians, another related people who came off the steppe to found an empire in what had been Persian territory" ([13]: 33). See also [12] and [4] for general histories of this formative period in the rise of Asian steppe pastoral nomad societies.

Simulation models of nomadic pastoralist societies include agent-based models of target systems ranging from regions in Africa (see, e.g., [2,18,25]) to the Arctic regions [3], the MASON Wetlands model [8], and a model of contemporary pastoralist behavior in Kazakhstan [21]. The model described in this paper builds on selected features of earlier models, by adding arguably more explicit social attributes and dynamics, and is also part of a larger interdisciplinary collaborative project aimed at investigating long-term adaptation and sociopolitical change in Inner Asia [9,10,23].

The emergence of sociopolitical complexity is a research question that models in our project seek to address. A new model was necessary because earlier models addressed different questions (e.g., sustainability, role of memory, or ethnic conflict). Specifically, the primary purpose of the HouseholdsWorld model is to gain a better understanding of pastoral nomadic dynamics – among households and between households and natural environments – including human and societal adaptive behavior to long-term change. For example, the HouseholdsWorld model is capable of answering research questions concerning societal consequences of climate change and other environmental challenges to human societies [24]. The second purpose is to provide a basis or building block for modeling a larger and more complex target system, including the emergence of political organization [5–7,10]. This is the paper-of-record for the MASON HouseholdsWorld Model version 1 (or HouseholdsWorld, for short).

2 The HouseholdsWorld Model

By way of context, the MASON HouseholdsWorld model is the initial social simulation model in a progression of models aimed at simulating the rise and fall of polities in Inner Asia over a long time span [9]. The time is defined as sufficiently long to include significant climate change. When climate changes, the biomass distribution on the landscape changes, which in turn leads to changes in the biological and social dynamics of animals and people, respectively.

HouseholdsWorld is a spatial agent-based model of pastoral nomads living in a simple socio-natural system, as shown by the UML class diagram in Fig. 1. The main agent classes are Household and Camp, where the former also belong to a Clan. The model is written in the MASON system [19] in order to exploit a set of project-specific features, such as platform independence in Java, guaranteed replicability, separation of computation from visualization, and evolutionary computation facilities which we plan to use in the future. Separate social, environmental, dynamic, and implementation aspects of the model are described next.

The target system is a generic locality smaller than a region of Inner Asia shortly after ca. 500 BCE, the time period just prior to the rise of the Xiung-nu polity (ca. 200 BCE). The primary sources used for developing the HouseholdsWorld model were epigraphic (e.g., [26]), archaeological, ethnographic, and environmental, as detailed in the subsections below. Secondary historical sources (e.g., [4,12,13]; and others), as well as area experts (see below), where also consulted.

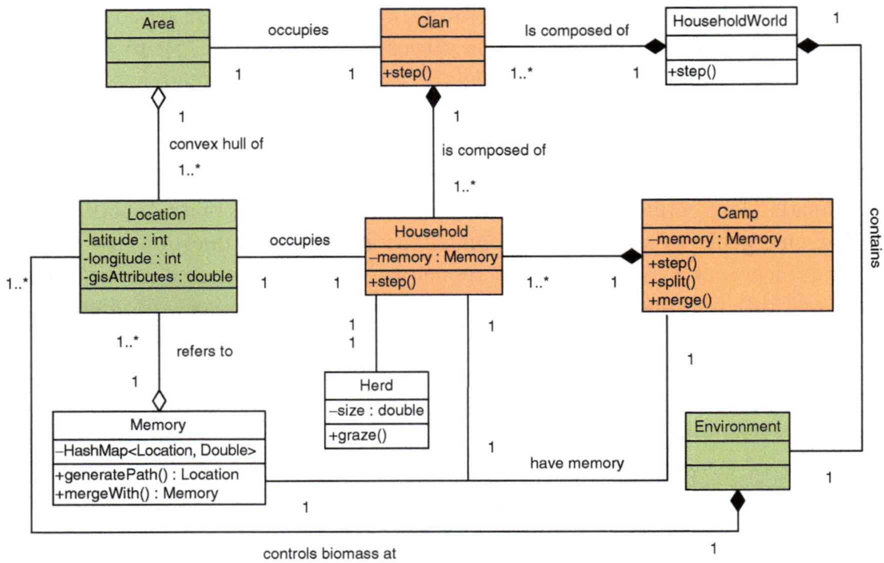

Fig. 1 High-level UML class diagram of the main components and relations in the Households World model, including the main attributes of Households and Camps. Agent classes (*orange*) and spatial classes (*green*) inherit from the MASON Steppable interface and from a subset of Geotools GIS attributes (describe areas, location), respectively

2.1 Households

Households are the smallest social units in the model. For our purposes, we do not specify individual persons, but rather focus on households – extended families – and their behavior around herds of animals. Each household needs to ensure sufficient forage for its herd (a herd multiplies if enough forage is available, otherwise it starts dying at a predetermined rate), as detailed by the main simulation loop illustrated in Fig. 2.

The movement rules mentioned in Fig. 2 are defined as follows. Each movement rule $r \in S$ takes as input a cell of the environment, $c \in N$, and returns a real number,

$$r : N \rightarrow \Re$$

The following four rules are currently used:[2]

[2] We are grateful to W. Honeychurch, W. Fitzhugh, B. Frohlich, and D. Tseveendorj for their expert advice on formulating these rules, based on the anthropological archaeological record of early Inner Asia and the ethnography reported for modern Mongolia.

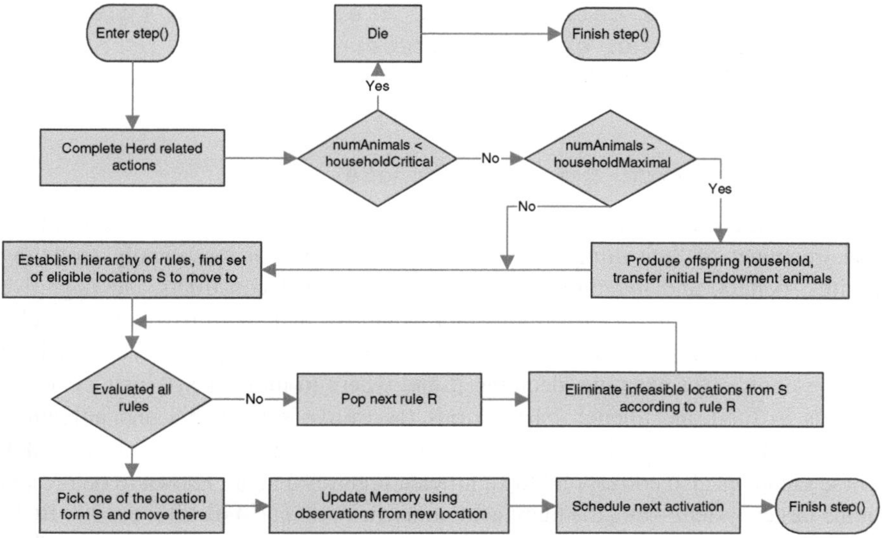

Fig. 2 Main simulation loop for HouseholdsWold agents: Flowchart detailing the process that takes place in the step method of the Household class. The loop starts at the upper left and ends at the bottom right

1. *Search for forage (local)*: Returns grass availability for an input cell. This is the primary rule for modeling household subsistence.
2. *Search for forage (global)*: Returns the reciprocal of distance from an input cell to the best grazing area, for a given time present in camp memory, using memorized abundance values of grazing for a given season of the year.
3. *Maintain camp cohesion*: Returns the reciprocal of the sum of Hamiltonian distances (defined as the minimal-length paths between cells) to camp members from an input cell.
4. *Avoid other camp members* and 5. *Avoid other clan members*: These return the sum of Hamiltonian distances from an input cell to all aliens (different clan/camp) present within a given Hamiltonian radius.

This rule set is defined at the level of households. The next section discusses the camp-level rule set.

The algorithm below shows how, given an ordered set of movement rules, agents decide where to move next. Note that Rule 2 requires endowing each household with grazing memory, which holds information on the best grazing areas for a given month of the year. This is a somewhat more sophisticated cognitive feature relative to our earlier Wetlands model where the hunter-gatherer agents followed a simpler rule set in a single ecological region [8].

Algorithm for evaluation of hierarchies of movement rules.

Given an ordered set of movement rules S and a set of eligible cells E **do**:
for all $r \in S$ **do**

```
  Evaluate each cell from set E according to rule r.
  Find the median evaluation score.
  Remove from E all cells falling below the median eval-
uation score.
end for
Draw a cell from E and move there.
```

While the social ontology of the HouseholdsWold model does not include elaborate treatment of "the mind of agents" (e.g., in terms of specifically modeling desires, beliefs, and intentions (DBI), or other mental constructs and cognitive processes), the model does include some cognitive and socio-psychological elements, such as agent memory (for both households and camps), decision-making by households (for example, deciding if and where to move next), instincts (with regards to food and aliens), basic norms (ascriptive xenophobia and attraction, determined by clan membership), and intentional behaviors (motivated by needs). These elements also correspond to significant features that are known to be present in the target system – i.e., the human communities in Inner Asia after ca. 500 BCE. Note also that, while rules 1 and 2 pertain to socio-natural interactions, rules 3, 4, and 5 are about interactions among households, consistent with a complex adaptive systems approach to the human ecology of small groups [1].[3]

2.2 Camps

Camps are composed of households, as shown earlier in Fig. 1, and they follow a separate set of behavioral rules:

1. *Division*: If households in a camp lack sufficient forage, the camp may divide into two new camps of approximately equal size. Division occurs along either north-south or east-west directions, with new camps drifting in opposite directions.
2. *Merging*: If two camps meet, each with enough supplies, they merge forming a new camp.[4]

Iterative application of these rules yields the effect of seasonal variability in the number of camps. Thus, in winter we observe multiple small camps, while in summer camps tend to conglomerate. In addition to behavioral rules, households belonging to a single camp share information on grazing areas.

[3]In the target system, the clan or tribal membership of households is far more consequential. For example, such associations can regulate patterns of conflict by segmentary or complementary opposition, a social feature we investigate in a different albeit related model in this project [10]. In the HouseholdWold model the clan or tribal membership of households only affects their camping behavior.

[4]In the target system the range of camp size is 5–8 households, with larger or smaller camps increasingly infrequent (improbable).

Importantly, camp behavior rules aim to reflect comparable rules in the target system of Inner Asia's pastoralist communities, similar to the motivation for household rules.

2.3 Natural Environment

The seasonal and spatial variability of the environment of Inner Asia, centered on Mongolia, is modeled by using two data layers:

1. *Monthly NDVI* (normalized difference vegetation index) rasters, from atmosphere corrected bands with a 500 m resolution; and
2. *Land cover type* (14 types used), originally in 1 km resolution.

To obtain carrying capacities from raw NDVI data, information gathered in situ by Kawamura [14] was used to translate NDVI into the amount of biomass fit for grazing (expressed in kg/ha). Parameter values are shown in Table 1. In all, the natural environment in the model consists of three biomes – desert, grasslands (mostly), and forests – representing a significant portion of the habitat occupied by the pastoral nomads of Inner Asia.

3 Simulated Dynamics

Like other models in MASON, HouseholdsWorld offers numerous data collection facilities, including social network and spatial clustering statistics, as well as an easy-to-use GUI for demonstrating experiments. (In this paper we omit MASON facilities relevant for evolutionary computation or related tools such as ECJ; see [20].) Figure 3

Table 1 Main parameters of the MASON HouseholdsWorld model

Parameter	Value	Meaning
`householdMaximal`	40	Maximal number of animals a household can support (if more, offspring will be produced)
`householdCritical`	5	Minimal number of animals required for the survival of a household
`householdVisionRange`	3	Vision range for a household
`InitialAnimalEndowment`	20	Number of animals that will be transferred to an offspring household
`initialNumberHouseholds`	100	Initial number of households set to a default value of 100
`biomassRegrowthRate`	1.5%	How fast does the biomass grow to levels for a given month per day

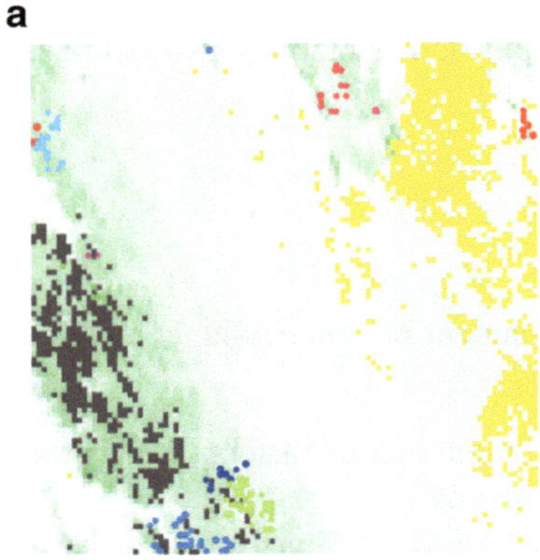

Map of households (by red, blue, green clans) and landscape.

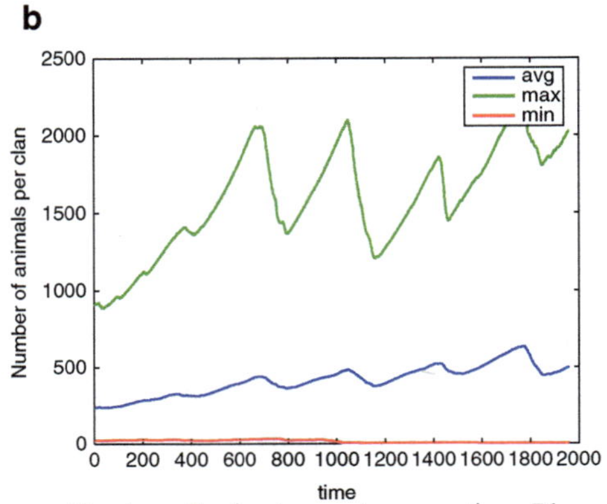

Number of animals per clan over time. Blue = average, Green = maximum, Red = minimum.

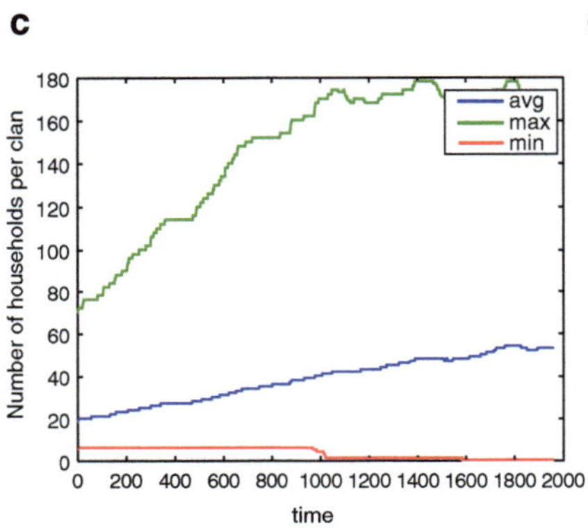

Households per clan over time. Legend as in (b).

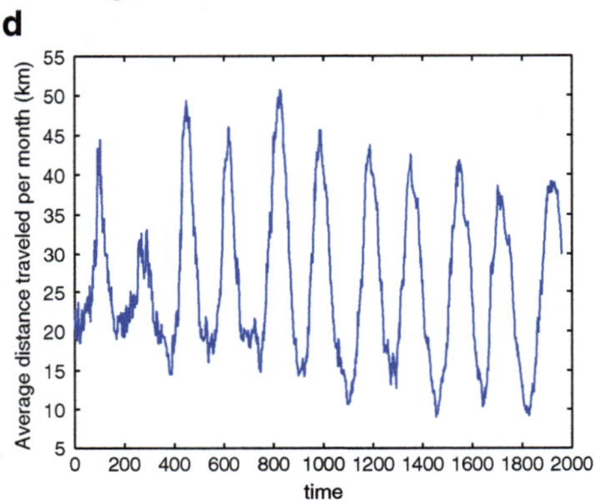

Average distance/month traveled by households.

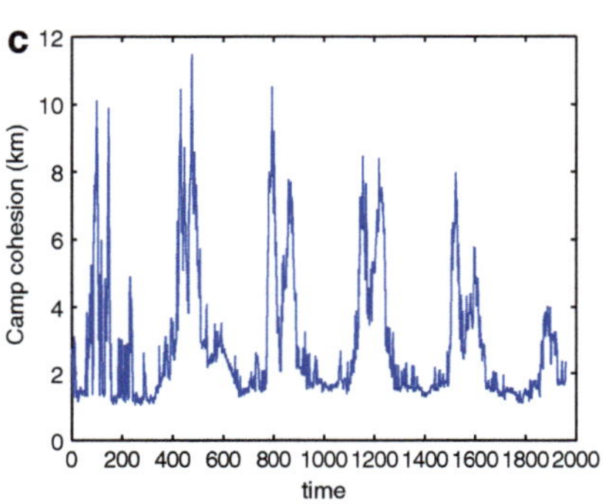

Cohesion (spatial aggregation) of camps over time.

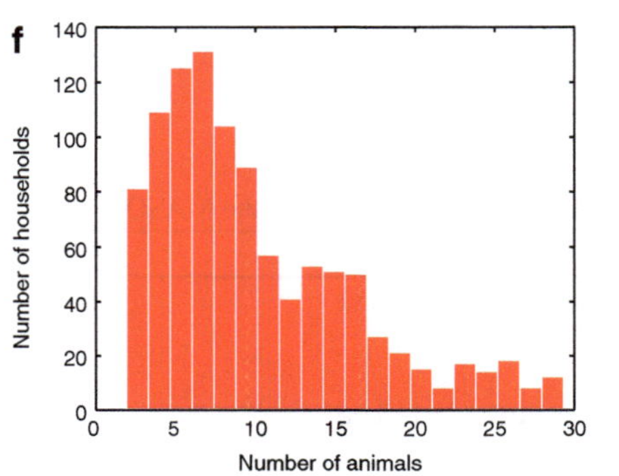

Simulated distribution of household wealth approximating a log-normal distribution.

Fig. 3 Examples of simulated outputs and statistics from a single run of the HouseholdsWorld model in default landscape of 100×100 km during summer. Timescales are expressed in days, with 300 simulated days per simulated year and approximately 6 years in each run. *Green* = biomass density, *Yellow* = desert, *Black* = forest

presents six selected outputs produced during each run of HouseholdsWorld. The model GUI displays a bird's eye view of a region where nomadic households form camps (clusters of red, blue, and yellow dots, representing three clans in panel (a)), as well as a variety of metrics (time series, histograms, and other charts not shown here for reasons of space) that track the evolution of social and/or environmental dynamics within the overall socio-natural system.

Several of the patterns produced by simulation bear significant qualitative and quantitative resemblance to comparable patterns in the target system. For example, the distribution of wealth in Fig. 3f has the approximate form of a log-normal distribution, as a real-world distribution of household wealth usually should. Similarly, household movements in Fig. 3d show marked periodic fluctuations, as in the real world when nomads undergo seasonal travel following their herds. While the model does not attempt to produce a specific historical or empirically replicated replication (such as, for example, migrations and settlement patterns in the well-known Anasazi model), the overall qualitative and quantitative behavior of households, herds, and seasons are supported by known features of the target system.

The MASON system of the HouseholdsWorld model also allows for easy design of multi-run experiments, as should be with a viable virtual lab (agent-based model). Accordingly, we have used the multi-run experimental capability to search for robust grazing strategies by households. For example, we have investigated performance across different permutations of the movement rule set.

Figure 4 presents results of an exhaustive search of the space of behavioral rule-set permutations to obtain the trade-off between the maximal long-term population and the long-term variability of the population. The Pareto frontier in this case is the set of solutions that can sustain the largest population within the larger set (i.e., as a subset) of all solutions that have comparable or larger risk measures (starvation probability). It can be observed that for populations less than 3,000, the tradeoff between risk and efficiency is small (population can increase with little punishment). However, the tradeoff changes dramatically for larger populations. For Mongolia, [17] provides observed population densities of 0.9 persons/km^2 and herd densities of 8 animals/km^2. For the particular landscape used in our simulation, the critical density is in the 3,000–5,000 households range on a 10,000 km^2 landscape (corresponding to a population density of 0.3–0.5 households/km^2 and a herd density of approximately 12–20 animals/km^2). Empirically observed values are some 50% lower than critical values predicted by the HouseholdsWorld model. We hope to reduce this discrepancy by introducing a data-driven model of paleo-climate, thereby increasing the environmental uncertainty faced by Inner Asian households in the real world.

4 Summary

Computational modeling of nomadic pastoralist societies that range over diverse natural environments poses interesting challenges, because household mobility affects the fabric of social relations (interactions among households) as well as socionatural

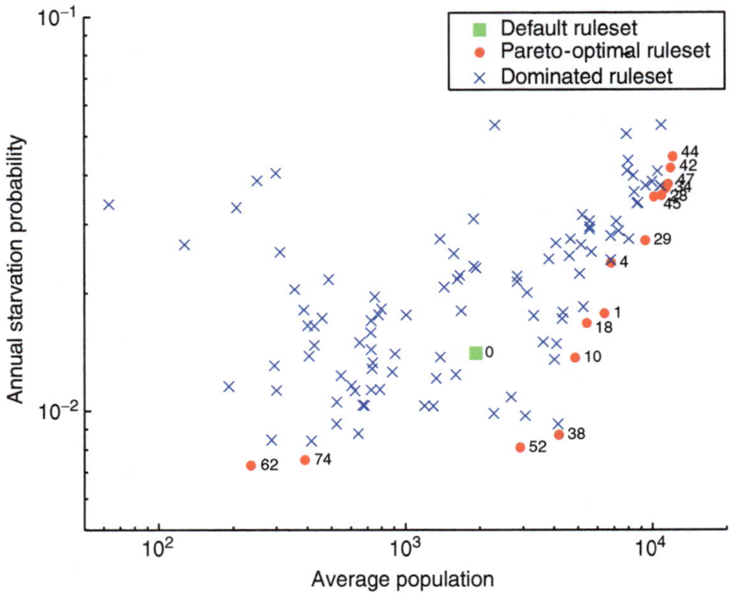

Fig. 4 Pareto frontier of permutations of the household behavioral rule-set in an average long term population plotted against starvation risk space. Note that the axes are in log-log scales. Each point represents the performance of the population following one of 120 (or 5!) possible orderings of rule sets. The permutation numbers from the graph correspond to lexicographic orderings of the default (0) rule set listed in Sect. 2.1. Legend: *Red*=points belonging to the Pareto-optimal rule-set frontier; *Blue*=dominated rule sets; *Green*=default rule set

interactions. In this chapter we have presented the MASON HouseholdsWorld, a new agent-based model of migratory pastoralist societies in a natural habitat consisting of several biomes (grasslands, forests, deserts). Agent-based models of pastoral nomadic societies are still relatively rare in the literature, although the role of these polities in world history has been significant [4, 12, 15, 22]. The MASON HouseholdsWorld model is part of a broader collaborative interdisciplinary project involving social scientists, computer scientists, and environmental scientists, focused on computational modeling of long-term human adaptations in Inner Asia – a vast area of the world where agro-pastoralist societies interacted among themselves and with their neighbors in complex and dynamic environments.

In this paper we described the structure and some of the dynamics of the HouseholdsWorld model, including fluctuations in household wealth, camp sizes, and other simulation outputs. The MASON HouseholdsWorld model is being used for conducting experiments on socio-environmental interactions and social dynamics, as well as for developing additional models that reach higher levels of social complexity (state-formation).

Acknowledgments An earlier version of this paper was presented at the second World Congress on Social Simulation, George Mason University, Fairfax, Virginia, USA, 14–17 July, 2008.

Funding for this study was provided by grant no. BCS–0527471 from the US National Science Foundation, Human and Social Dynamics Program, and by the Center for Social Complexity at George Mason University. The authors thank all members of the Mason-Smithsonian Joint Project on Inner Asia, especially Bill Honeychurch, Sean Luke, Bruno Frohlich, Bill Fitzhugh, Max Tsvetovat, and Dawn Parker, as well as three anonymous referees, and D. Tseveendorj of the Mongolian Academy of Sciences, Institute of Archaeology, Ulaanbaatar, Mongolia, for comments and discussions.

References

1. Arrow H, McGrath JE, Berdahl JL (2000) Small groups as complex systems. Sage Publications, Thousand Oaks, London, and New Delhi
2. Bah A, Toure I (2006) An agent-based model to understand the multiple uses of land and resources around drillings in the Sahel. Math Comput Model 44:513–534
3. Berman M, Nicholson C, Kofinas G, Tetlichi J, Martin S (2004) Adaptation and sustainability in a small Arctic community: results of an agent-based simulation model. Arctic 57:401–414
4. Christian D (1998) A history of Russia, Central Asia and Mongolia. Blackwell, Oxford, UK
5. Cioffi-Revilla C (2010) On the methodology of complex social simulations. J Artif Soc Soc Simulat 13(1) 7 <http://jasss.soc.surrey.ac.uk/13/1/7.html>
6. Cioffi-Revilla C (2009) Simplicity and reality in computational modeling of politics. Comput Math Organ Theor 15(1):26–46
7. Cioffi-Revilla CS, Paus SL, Olds JL, Thomas J (2004) Mnemonic structure and sociality: a computational agent-based simulation model. In: Sallach D, Macal C (eds) Proceedings of the agent 2004 conference on social dynamics: interaction, reflexivity and emergence. Argonne National Laboratory and University of Chicago, Chicago
8. Cioffi-Revilla C, Paus S, Luke S, Olds JL, Thomas J (2004) Mnemonic structure and sociality: a computational agent-based simulation model. In: Sallach D, Macal C (eds) Proceedings of the agent 2004 conference on social dynamics: interaction, reflexivity and emergence. Argonne National Laboratory and University of Chicago, Chicago, IL
9. Cioffi-Revilla C, Luke S, Parker DC, Rogers JD, Fitzhugh WW, Honeychurch W, Frohlich B, DePriest P, Amartuvshin C (2007) Agent-based modeling simulation of social adaptation and long-term change in Inner Asia. In: Takahashi S, Sallach D, Rouchier J (eds) Advancing social simulation: the first world congress. Springer, New York and Tokyo
10. Cioffi-Revilla C, Honeychurch W, Latek M, Tsvetovat M (2008) The MASON Hierarchies model of political organization and warfare: paper of record. Technical report, Mason-Smithsonian Joint NSF/HSD Project on Inner Asia
11. Epstein JM (2007) Generative social science: studies in agent-based computational modeling. Princeton University Press, Princeton, NJ
12. Grousset R (1970) The Empire of the Steppes: a history of Central Asia. Rutgers University Press, New Brunswick, NJ and London
13. Hildinger E (1997) Warriors of the steppe: a military history of Central Asia, 500 B.C. to 1700 A.D. Da Capo Press, Cambridge, MA
14. Kawamura K, Akiyama T, Yokota H (2004) Comparing modis vegetation indices with avhrr ndvi for monitoring the forage quantity and quality in Inner Mongolia grassland and China. Japanese Society of Grassland Science
15. Khazanov AM (2004) Nomads of the Eurasian steppes in historical retrospective. In: Grinin LE, Carneiro RL, Bondarenko DM, Kradin NN, Korotayev AV (eds) The early state, its alternatives and analogues. Uchitel Publishing House, Volgograd, Russia, pp 476–500
16. Kohler TA, van der Leeuw S (2007) Model-based archaeology of socionatural systems. School of American Research (SAR) Press, Santa Fe, NM

17. Krader L (1955) Ecology of Central Asian pastoralism. SW J Anthropol 11:301–326
18. Kuznar LA, Sedlmeyer R (2005) Collective violence in Darfur: an agent-based model of pastoral nomad/sedentary peasant interaction. Math Anthropol Cult Theor 1. Available at: http://www.mathematicalanthropology.org/pdf/KuznarSedlmeyer1005.pdf
19. Luke S, Cioffi-Revilla C, Panait L, Sullivan K, Balan GC (2005) MASON: a multiagent simulation environment. Simulation 81:517–525. Available at: http://cs.gmu.edu/~eclab/projects/mason/
20. Luke S, Panait L, Balan G, Paus S, Skolicki Z, Popovici E, Sullivan K, Harrison J, Bassett J, Hubley R, Chircop A (2009) ECJ 18: a Java-based evolutionary computation research system. Available at: http://www.cs.gmu.edu/~eclab/projects/ecj/
21. Milner-Gulland EJ, Kerven C, Behnke R, Wright IA, Smailov A (2006) A multi-agent system model of pastoralist behaviour in Kazakhstan. Ecol Complex 3:23–36
22. Rogers JD (2007) The contingences of state formation in Eastern Inner Asia. Asian Perspect 46:249–274
23. Rogers JD, Cioffi-Revilla C (2010) Expanding empires and the analysis of change. In Current Archaeological Research in Mongolia, edited by J. Bemann, H. Parzinger, E. Pohl and D. Tseveendorzh. Bonn, Germany: Bonn University Press.
24. Rogers JD, Latek M, Nichols T, Emmerich T (2009) Weather, scale, and complexity in Inner Asian pastoralist adaptive strategies. Working paper, Washington, DC. Mason-Smithsonian Joint Project on Inner Asia.
25. Rouchier J, Bousquet F (2001) A multi-agent model for transhumance in North Cameroon. J Econ Dynam Contr 25:527–559
26. Watson B (ed) (1961) Records of the Grand Historian of China. Columbia University Press, New York

Effects of Adding a Simple Rule to a Reactive Simulation

Pablo Lucas

Abstract This paper focuses on simple affective roles in a replication of the Dominance World (DomWorld) model of primate social behaviour. Agents are discussed as autonomous entities capable of managing their social ranks by performing or avoiding dominance interactions. With autonomy described as the ability to deal with such aggressions only by using local perception and internal action-selection, different social organisations can be observed by introducing a simple, reactive representation of fear to some agents in the model.

Keywords Agent architecture • Social organisation • Reactive action-selection

1 Introduction

Most agent-based social simulation (ABSS) architectures combine sensor-based and reactive behavioural rules. In the primate social simulation context, however, ethological and ecological models are commonly developed focusing using approaches similar to game theory. That is, with particular attention to optimising certain individual capabilities, which are usually interpreted as subject to adaptation by natural selection – for example survival, reproduction, hunting and foraging skills [1].

Despite the importance of maximising efficiency of resource usage and individual fitness, these are not the only drives of primate social interactions. Distribution of resources, relationship dynamics and other environmental conditions are known to affect both group organisation and individual traits [2]. But how can affective concepts be applicable within the context of an ABSS? Emotional properties can be implemented to interfere both at symbolic and non-symbolic components of the agent architecture. Thus, agent-architectural recommendations depend on what such entities are

P.L ucas(✉)
CentreF orPol icyM odelling, Manchester,M 13G H,Engl and,U K
e-mail:pa blo@cfpm.org

K. Takadama et al. (eds.), *Simulating Interacting Agents and Social Phenomena: The Second World Congress*, Agent-Based Social Systems 7,
DOI 10.1007/978-4-431-99781-8_15, © Springer 2010

expected to perform at runtime. Although primate individual aggressions (dominance interactions) and conflict avoidance are commonly analysed using ethological cognitive strategies of exchange, the Dominance World (DomWorld) model, originally implemented in Object Pascal [3], is an example that can generate plausible data regarding empirical observations using only cycles of reactive behavioural rules.

2 Extending the Original Model

Although there are numerous approaches to model social primate behaviour [1], the DomWorld [3] and its precursor framework MIRROR [4, 5] is one of the most analysed simulations of plausible rank-based social organisation through aggressive interactions. The original DomWorld model focused on homogeneous agents, distributed on a grid, executing differentiated male and female pre-fixed rules. These include a one-unit forward move, direction rotations and interactions with other agents occupying adjacent cells. The rank dynamics is dependent on each agent internal setting, as these will influence the intensity and frequency of interactions.

a

$$w_i = \begin{bmatrix} 1 & \frac{Dom_i}{Dom_i + Dom_j} > Random(0{,}1) \\ 0 & else \end{bmatrix}$$

b

$$Dom_i = Dom_i + \left(w_i - \frac{Dom_i}{Dom_i + Dom_j}\right) * StepDom \quad (2)$$

$$Dom_j = Dom_j + \left(w_i - \frac{Dom_i}{Dom_i + Dom_j}\right) * StepDom$$

The only exception to the above equations is that the lowest

Fig. 1 (**a**) A dominance interaction, and (**b**) rank adjustments [3][1]

With 1 indicating success and 0 defeat, Fig. 1a illustrate the mechanics of aggressions considering ranks (Dom) of involved agents *j* and *i*. According to Fig. 1b, winners will have their rank increased whilst moving towards the rival. On the other hand, losers decrease their ranks and move two units away using a random angle between 0 and 45. Winning probability of dominance interactions depends of current ranks (Dom) and aggression intensities (StepDom percentages) of involved agents, letting those on top of the social hierarchy with greater changes of improving ranks.

The simulation uses configurable parameters of individual reactivity listed in Table 1. Due to natural differences in size and strength, DomWorld initialises male

Table 1 Configurable parameters in the simulation adapted for this paper

Parameter	Possible values	Description
Population	Minimum 2, maximum 40	Number of agents
Aggression	Low (1%) to high (100%)	Intensity of an attack
Attraction	Boolean (true or false)	Males look for females
Epoch	Range from 1 to 1,000	When to log simulation data
Cycles	Range from 1 to 200	Simulations per epoch
Fear	Boolean (true or false)	Female avoidance of males

[1]All values are updated recursively and negative ones are avoided by a minimum rank of 0.01.

ranks as 16, whilst females start with eight and have 80% of the male aggression intensity [3]. This version of the model only differs from the original by not calculating centrality measurements and introducing reactive fear as an individual interference to usual female behaviour during their period of tumescence. Figure 2 shows simulation cycles and behavioural rules in the model, including fear implemented as heading females away (by choosing a random angle between 0 and 45) from the nearest visible male agent within 24 units. Double lined polygons indicate use of pseudo-random parameters, seeded with a 32,000 integer. The main action-selection loop consists of each agent initially checking if its personal space of 2 cells was invaded. If not, dominance or grouping behavioural rules are processed sequentially using an unordered list of agents provided at runtime by the implemented simulation.

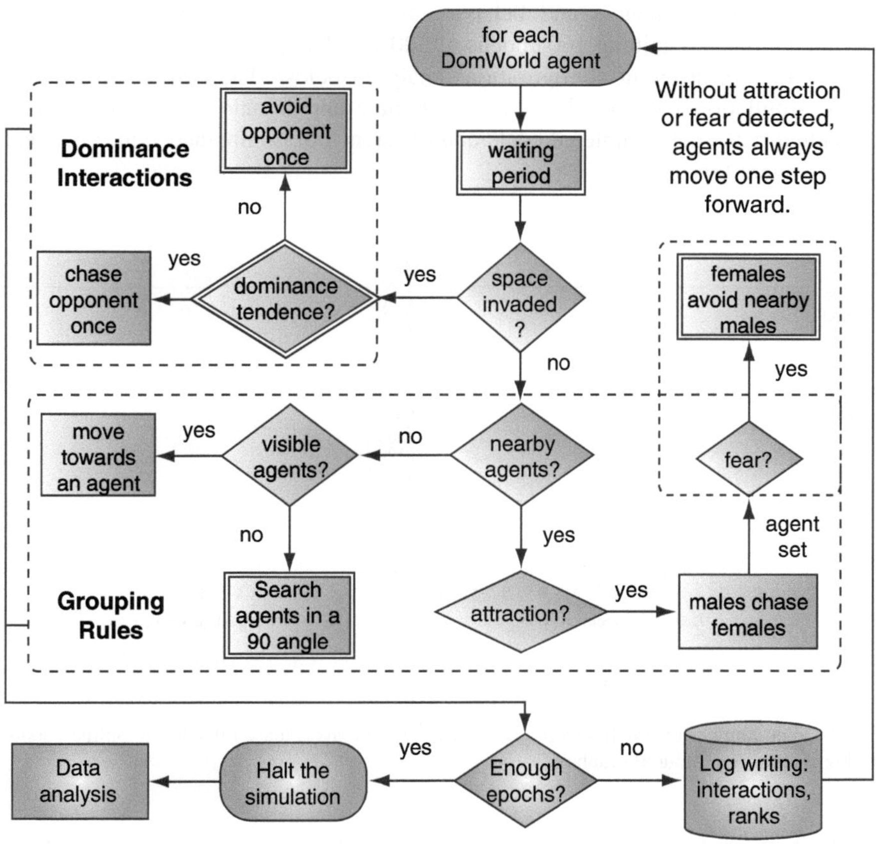

Fig. 2 Simulation cycles

It is known that primates may deliberately avoid invading the personal space of other to avoid triggering unnecessary conflicts [6, 7]. But when there is too much proximity, dyadic dominance interactions are likely to happen amongst the nearest ones in aggressive societies. In the simulation model, if no dominance interaction takes place, agents check for other ones in their nearby configured vision of 24 units. If negative, due to their hard-coded drive for grouping, they move 1 unit towards any observed agent in their maximum vision of 50 units. Alternatively, a 90 degrees search can be executed in case no other entity could be sensed. If attraction is enabled, males try to re-orientate themselves toward females in their nearby view; whilst in case of fear, females try to keep themselves away from any male agent.

3 Social Organisation

Using original parameters and behavioural rules, ten consecutive runs of the extended model containing 160 epochs of 200 simulation cycles were executed for each of the six different configurations analysed below. The output for all completed simulation includes the number of male and female driven interactions, individual ranks per completed cycle and difference of dominance values between the two agent types.

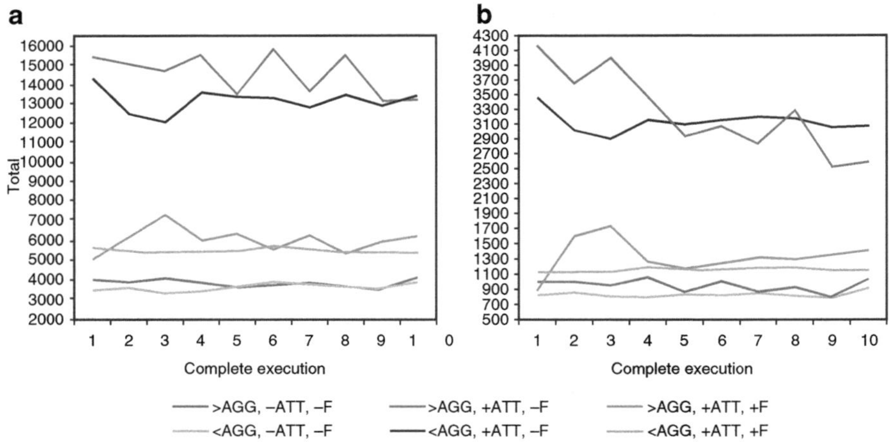

Fig. 3 Male number of: (**a**) interactions, and (**b**) aggressions (Please refer to the online version of this chapter for coloured graphs)

Figure 3a displays the total number of male interactions during all executions with different configurations, and Fig. 3b depicts the number of male aggressive interactions. Enabling attraction (+ATT) without fear (–F) results in a considerable increase of both interaction types. This effect tends to increase the overall number of male interactions in highly aggressive conditions (>AGG) in Fig. 3a, yet it can also lead to decrease the total dominance interactions when aggression levels are low (<AGG) as in Fig. 3b. This characteristic is directly related to male rank values in >AGG conditions. A further discussion about this note is made while addressing Fig. 5.

When female fear is activated (+F) in cases of both high and low aggression levels, the number of interactions shown in Fig. 3a and b are lower than without fear (−F). Whilst total male interactions and aggressions are high throughout the simulations with enabled attraction (+ATT), simulating agents without fear (−F) increases their chances of interacting between themselves as no female would attempt to avoid males approaching their nearby personal space of two units. Although in real primate societies fearful behaviour can also be observed with the introduction of unfamiliar objects or animals to their natural habitat [2, 7], the reactive extension made for this paper only focused on female fear to simply test the effect of a hypothetical, reactive learning.

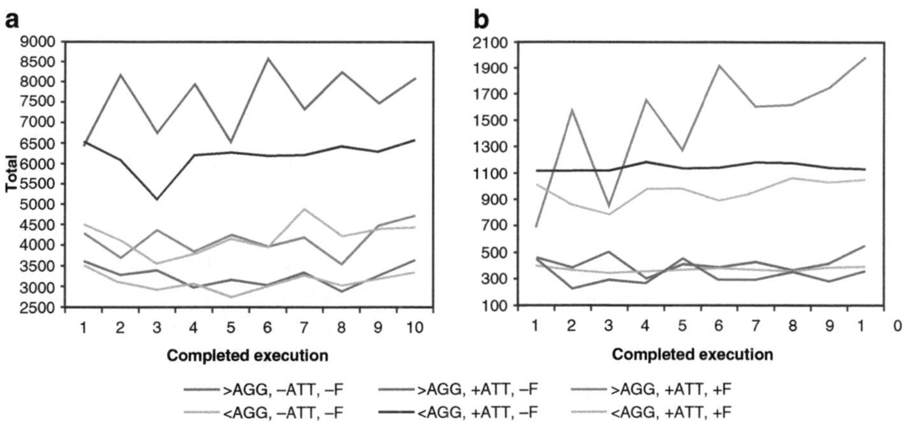

Fig. 4 Female number of: (**a**) interactions, and (**b**) aggressions (Please refer to the online version of this chapter for coloured graphs)

In Fig. 4, the total amount of interactions tends to increase in cases of enabled attraction (+ATT), both with high (>AGG) and low (<AGG) aggression levels. When fear is introduced in a highly aggressive configuration with attraction (>AGG, +ATT, +F), female agents perform more dominance interactions than in simulations with high aggression and attraction but without fear (>AGG, +ATT, −F). Conversely, the total number of non-aggressive interactions tends to decrease under the same conditions enabling high aggression, attraction and fear (>AGG, +ATT, +F).

It can be noted in Figs. 3 and 4 that the total number of male interactions and aggressions is higher than female in every simulation configured with high levels of aggression. Attraction significantly increases the total number of interactions in any simulation. Whilst fear tends to reduce both male totals, the frequency of female attacks is likely to increase. No female agent presented higher rank than males in configurations with low aggression, as all their dominance values were

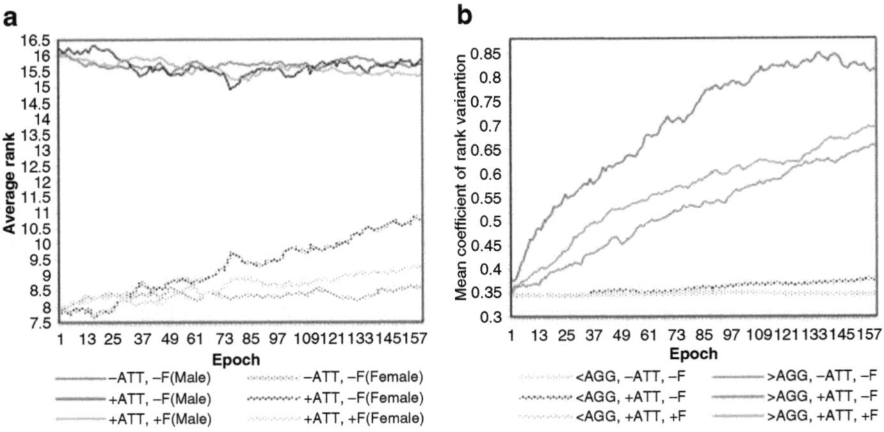

Fig. 5 (**a**) Ranks with >AGG, and (**b**) Rank variation [7] (Please refer to the online version of this chapter for coloured graphs)

kept between 7.5 and 8.5, whilst male ranks varied from 15.5 to 16.5. More aggressive interactions naturally favour males due to their more intense attack power and higher initial ranks.

Figure 5a shows how male and female average ranks evolve over the simulated runs in different configurations of high aggression; while Fig. 5b indicates the average variation between ranks. Higher values represent a steeper social hierarchy. Occasionally some females present higher ranks than males. However, the non-dashed lines show the continuous supremacy of male ranks over the female values (dashed lines). The figure shows female ranks tending to improve most when the simulation is configured with attraction and without fear (+ATT, –F), while Fig. 5b shows the coefficient of rank differentiation is practically stable in these conditions.

Nevertheless, when fear is introduced in highly aggressive configurations with attraction, (Fig. 5b) >AGG, +ATT, +F, overall females tend to have better ranks than without aggressive attraction. The associated coefficient has much smaller values than with high aggression levels only with attraction, (Fig. 5b) >AGG, +ATT, –F, and expectedly somewhat higher than in simulations configured as highly aggressive, without attraction or fear. This extended model may occasionally favour or impair the ability of agents to maintain higher ranks. This feature is better observed amongst female ranks, as in configurations with attraction, together with fear and intense aggression, males engage in more aggressive interactions amongst themselves whilst conflicts between genders are less frequent than without fear. When fear is introduced, the difference between male and female ranks can decrease simply by interfering in the pattern of aggressive interactions. This exemplifies how a simple, reactive rule can play important roles in changing the results obtained in a social simulation model.

4 Final Considerations

This paper includes a discussion on the impact of introducing reactive fear into a social model of primate social organisation. Emotion is discussed as a plausible influence in the execution of local (individual) action-selection mechanisms and shown how it can substantially alter the results of simulations over time. Running conflicting male and female action-selection procedures suggests system-wide effects similar to attractors found in system dynamics models [8]. That is, the effect of local (individual) and periodical interference that alter global properties of a closed system. This include, for example, results of the extended model providing different social organisations that are processed in a completely decentralised form, as the dynamics of dyadic interactions occurring at local levels are the only way to influence results.

Acknowledgments The author would like to acknowledge helpful discussions with Charlotte Hemelrijk, Ruth Aylett, David Wolfe Corne and Bruce Edmonds and support from ETH Zurich, team of Dirk Helbing (Chair of Sociology, in particular of Modeling and Simulation).

References

1. Dunbar RIM (2002) Modeling primate behavioural ecology. Int J Primatol 24:4
2. Huffman MA (1996) Acquisition of innovative cultural behaviours in nonhuman primates: a case study of stone handling, a socially transmitted behaviour in Japanese macaques. In: Heyes CM, Galef BG Jr (eds) Social learning in animals. Academic, San Diego, pp 267–289
3. Hemelrijk CK (1999) An individual-oriented model of the emergence of despotic and egalitarian societies. Proc R Soc Lond 266:361–369
4. Hogeweg P, Hesper B (1985) Socioinformatic processes: MIRROR modelling methodology. J Theor Biol 113:311–330
5. te Boekhorst IJA, Hogeweg P (1994) Self-structuring in artificial CHIMPs offers new hypotheses for male grouping in chimpanzees. Behaviour 130:229–252
6. Flack JC, de Waal FBM (2004) Dominance style, social power, and conflict management in macaque societies: a conceptual framework. In: Thierry B, Singh M, Kaumanns W (eds) Macaque societies: a model for the study of social organization. Cambridge University Press, England, pp 157–181
7. Pablo Lucas dos A, Ruth A, Alison C (2007) Adapting hierarchical social organisation by integrating fear into an agent architecture. In: 7th International virtual agents (IVA), Paris, France, September 17–19 2007. Springer
8. Sallach DL (2000) Classical social processes: attractor and computational models. J Math Sociol 24:245–272

Applying Image Texture Measures to Describe Segregation in Agent-Based Modeling

Kathleen Pérez-López

Abstract Many agent-based models produce output as multi-region patterns on a two dimensional grid. The degree of mixing versus segregation in these displays is a particularly interesting outcome of some of these models, and comparing output patterns is an important component of evaluating their results. Typically these comparisons are made in an ad-hoc and qualitative manner. This paper proposes applying measures used to describe textures in the field of image processing to quantitatively evaluate patterns resulting from agent-based models. The rationale for using these measures is made by relating the underlying image formation process, based on Markov Chain Monte Carlo methods, to the Schelling segregation process. Examples are presented that support the premise that these statistics capture the degree of segregation in a 2D display. The ideas are also tested on the output of the agent based model, Wetlands (Cioffi-Revilla C, Paus S, Luke S, Olds JL, Thomas J (2004) Mnemonic structure and sociality: a computational agent-based model. In: Conference on collective intentionality IV, Certosa di Pontignano, Siena, Italy, 13–15 October 2004), supporting the premise.

Keywords Agent-based model • Image texture features • Schelling segregation • Gibbs random field

1 Introduction

Many agent-based models produce output as multi-region patterns on a two dimensional grid. The degree of mixing versus segregation is a particularly interesting outcome of some of these models, and comparing output patterns is an important component of evaluating their results. However, typically comparisons are made in an ad-hoc and qualitative manner.

K. Pérez-López (✉)
American Institutes for Research, 1000 Thomas Jefferson Street, NW, Washington, DC20007, USA
e-mail: kperez-lopez@air.org

K. Takadama et al. (eds.), *Simulating Interacting Agents and Social Phenomena: The Second World Congress*, Agent-Based Social Systems 7,
DOI 10.1007/978-4-431-99781-8_16,

Image processing researchers have used the Gibbs random field (GRF) to characterize complex natural textures in images since the 1980s [1, 2]. The Gibbs formulation, borrowed from statistical mechanics, enables the analysis of image patterns by providing a distributional form for them, treating them as if their pixels were interacting particles. Complementing this distributional form, Markov Chain Monte Carlo (MCMC) methods prescribe processes that, starting from random configurations of pixels, slowly rearrange them and eventually produce highly probable configurations.

In 1971 [3], prior to the application of the GRF in image processing, Thomas Schelling proposed a model of segregation that results in patterns remarkably similar to some of those produced by the image simulation methods. These visual similarities support the use of the image texture measures to describe segregation patterns.

This paper presents the connections between the image simulation and Schelling segregation processes and derives measures from the former that can be applied to the latter. Part 2 describes the GRF and how it is used in image simulation processes. Part 3 overviews the Schelling segregation model and points out the similarities to a particular GRF image model. Part 4 develops the image texture metrics based on the GRF process and provides examples that demonstrate how higher valued metrics correlate with increased clustering in the images. Part 5 describes some preliminary experimentation applying one of these metrics to the output of Wetlands [4], an agent based modeling system developed at George Mason University. Part 6 briefly mentions how the underlying processes described here are related to other processes involved in social simulation, including spatial potential games on graphs, and social networks. These relationships point to how the image texture measures are simple cases of parameters for these more complex embedding spaces. Part 7 concludes.

2 Gibbs Random Field in Image Processing

2.1 Definition

A GRF is a random field that is spatially Markov. That is, a GRF is a configuration of sites randomly assuming values within a specified range, and a neighborhood system that defines the conditional dependence of each site's value upon those of the other sites. Thus, for texture characterization in image processing, given an $N \times M$ lattice $L = \{(1,1)\ldots(1, M), \ldots,(N,1)\ldots(N,M)\}$, the random field would be the image X that assumes a particular configuration $x = \{x_{ij}\}$, where x_{ij} is the value at site (i, j) of L, taken from the set $Q = \{q_1, \ldots, q_m\}$. Here we consider Q to be the possible pixel colors or grey-levels, although in some applications it is a set of other characteristics, such as region assignments or edge designations [1]. The neighborhood

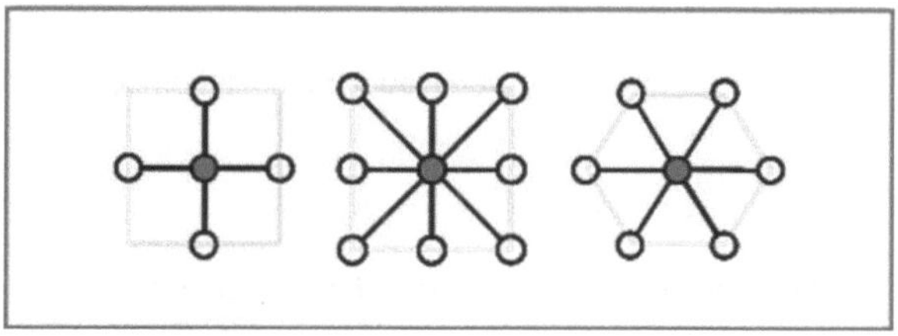

Fig. 1 Range 1 von Neumann, Moore and hex neighborhoods: white sites are neighbors of the central grey site

system, $\mathcal{N}$, defines a neighborhood N_{ij} for every site (i, j). Typical neighborhood systems are homogeneous across the image, and the N_{ij} are usually considered to be range 1 von Neumann, Moore or hex neighborhoods, as in Fig. 1.

Then the Markov property applied to this field states that the probability that a site assumes a value q in Q is independent of the values of any sites outside its neighborhood, as in (1).

$$P(x_{ij} = q \mid x_{kl} \in x) = P(x_{ij} = q \mid x_{kl} \in N_{ij}) \tag{1}$$

In 1971 Hammersley and Clifford proved a theorem that equates a random field satisfying the Markov property, plus the additional condition of positivity, with a Gibbs distribution [5]. Positivity requires that every configuration of values have a positive chance of occurring. Equations (2) through (4) present the Gibbs distribution over the possible configurations of the random field.

$$p(X = x) = \frac{1}{Z}\exp[-U(x)] \tag{2}$$

$$Z = \sum_{x} \exp[-U(x)] \tag{3}$$

Z is called the partition function, and used in the factor 1/Z; it normalizes the equation to sum to 1 over all possible configurations. The exponent in these equations, U(*x*), is the energy function; it is equal to the sum of *clique potentials*, $V_c(x)$, as seen in (4).

$$U(x) = \sum_{c \in C} V_c(x) \tag{4}$$

Clique potentials are arbitrary functions that depend only on the values of members of cliques, which are defined to be either the site itself, or sets of mutual neighbors as specified by the neighborhood system, $\mathcal{N}$. Figure 2 depicts all possible cliques for the range 1 von Neumann, Moore and hex neighborhoods. The parameters in the figure will be explained subsequently.

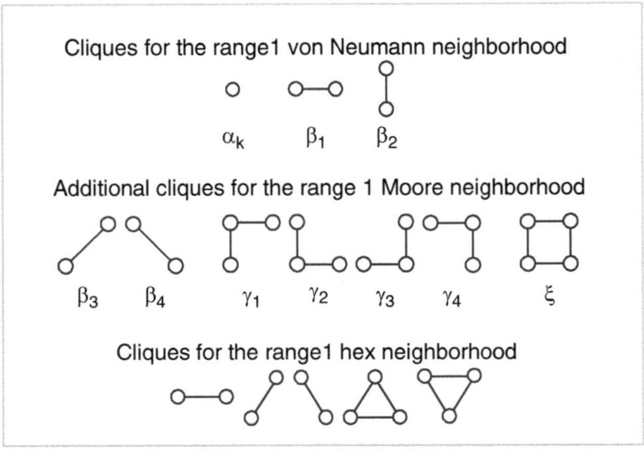

Fig. 2 Cliques defined by von Neumann, Moore and hex neighborhood Systems; labels below some clique represent associated parameters

Choosing a GRF model for a textured image then requires the following:

- A lattice of sites, L
- A set of possible site values, Q
- A neighborhood system, $\mathcal{N}$
- A potential function, $V_c()$, for each type of clique, c, and values for any associated parameters

2.2 *Image Simulation with Markov Chain Monte Carlo Methods*

Given a choice of an image model, a highly probable configuration can be simulated using MCMC methods to minimize the energy function. (Comparing the result to the source image, statistically and visually, can support the use of the chosen model.) Feller [6] proved in 1950 that a homogeneous, irreducible, aperiodic Markov chain has a unique limiting distribution. For a given target distribution, say the one defined by an image texture model under consideration, MCMC methods can produce such a chain X(t) of configurations: X^0, X^1, X^2, …, with that limiting distribution. There are different varieties of MCMC methods, but a general scheme to achieve the desired distribution in the form of (2) is shown in Fig. 3.

For all values of T, the chance of accepting a configuration as the next one in the chain is 1 if the energy decreases; otherwise, the chance of acceptance is still non-zero, but larger energy increases imply lower chances of acceptance. At the initial stages, the high value of T means that at step 1.c, the probability of accepting the altered configuration is very high, regardless of the amount of energy increase. This probability decreases until at very low T, virtually no increases in energy will be accepted. The starting value for T, the number of iterations at each value, and the rate at which it is reduced is called the cooling schedule. If the system is not hot

1. Start at a random configuration x^0 of site values chosen from Q and a very high value for the algorithm parameter, T
   a. At time t, in cycle k, tweak x^t slightly to get a candidate x^* for x^{t+1}
   b. Compute the change in the energy function, $\Delta = U(x^*) - U(x^t)$
   c. Let p = min[1, exp(-Δ/T)]; set $x^{t+1} = x^*$ with probability p, $x^{t+1} = x^t$ with probability 1-p
   d. Continue at (a) until Δ for the cycle stays "very low"
2. Reduce T. Repeat at (1.a) until T is "very low."

Fig. 3 An MCMC method to minimize the Gibbs energy function and achieve a mode of the target distribution

enough initially, or if cools too quickly, it will freeze, or quench, into a sub-optimal configuration. Van Laarhoven and Aarts [7] describe some criteria for the cooling schedule that guarantee a global minimum for U(x). These are very stringent criteria, and in practice they are only followed approximately, resulting in very good, but sub-optimal configurations.

2.3 *Example: An Ising/Potts Class of GRF*

In 1987, Derin and Elliot [2] published a particular implementation of the GRF based on a Potts model, an extension of the model proposed by Ising in 1925. Typically both the Ising and Potts models reside on a 2D rectangular lattice of sites. For the Ising model, Q, the set of possible site values, is $\{-1, 1\}$; for the Potts model Q may contain more than two values. Derin and Elliott use a range 1 Moore neighborhood with the Potts model. They define the clique potential functions to be homogeneous across the image, and to depend only upon the type of clique, as follows.

Each of the nine multi-member cliques for this neighborhood system is assigned a clique parameter, as shown in Fig. 2. The clique potential, $V_c(x)$, for clique c in configuration x is then determined by whether or not the members of the clique are all equal. Letting ζ represent the clique's assigned parameter, its potential function is:

$$V_c(x) = \begin{cases} -\zeta \; if\; all\; members\; of\; clique\; x\; are\; equal \\ \zeta \; otherwise \end{cases} \tag{5}$$

$$V_c(x) = a_k if \quad x_{ij} = q_k. \tag{6}$$

Using MCMC methods, Derin and Elliott produce highly probable configurations for distributions defined in this way, for various sets of parameter values. One of these configurations from their 1987 paper is the image in Fig. 4b. It is a three grey-level image, formed using parameter values of 1.0 for all two-member cliques

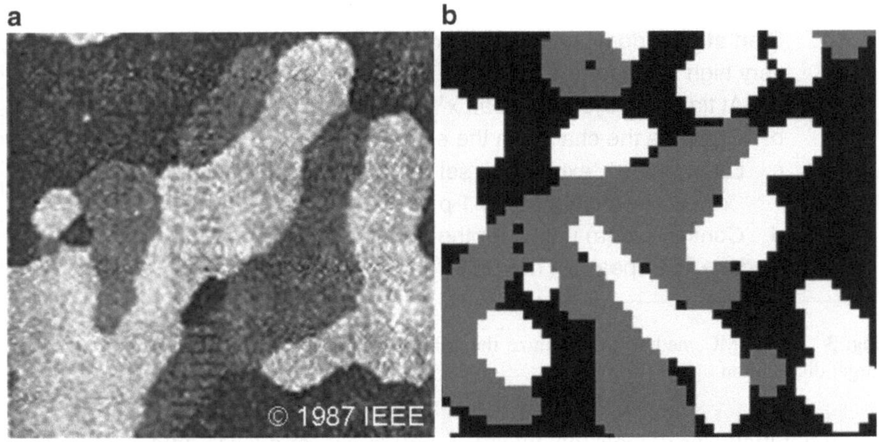

Fig. 4 (**a**) From Derin and Elliott [2], modes of Potts distributed image textures; (**b**) example of Schelling segregation process outcome, produced by the agent-based modeling program MASON [9]

and 0 for all three- and four-member cliques. The equal-valued two-member cliques lead to the isotropic patterns. The next section briefly describes the Schelling segregation process and shows how a slight variation of it can be expressed as the Derin and Elliott image formation process. The image in Fig. 4b shows the outcome of the Schelling process produced by the agent-based modeling tool, MASON [9]. The regions formed from both processes are similar in character.

3 Schelling Segregation Model

3.1 Original Schelling Model

In his segregation modeling process, Schelling demonstrated that individuals from two different groups, acting solely on the impulse to have some neighbors of their own group, would cause a system to become highly segregated, even when individuals had absolutely no aversion to members of the other group [3]. As Schelling described this process in 1978 [10], it was played on an 8×8 grid penciled on a sheet of paper. Dimes and pennies representing individuals in a community were randomly placed on the squares. An individual's neighborhood was represented by the eight nearest surrounding coins, except for coins on the boundaries, which had truncated neighborhoods. A happiness threshold described the proportion of neighbors of one's own type one would need to feel happy. At each play of the game, an unhappy individual was randomly selected. If there were empty squares in which the individual would be happier, the coin was moved to the closest of them. Play continued until each coin represented a happy individual, or there were no empty spaces to which an unhappy individual could move to be happier. Games vary by the number of sites occupied by each group, (which could leave some spaces empty), and by the happiness threshold. High happiness thresholds produced strong segregation patterns, as one would expect. But surprisingly, many games with only a slight preference for like neighbors ended in highly segregated patterns of coins.

Our conjecture is that the preference for like neighbors can be captured by a Gibbs energy function with Derin & Elliott clique parameters, and that the game play can be described as an MCMC process starting with T so low that the system is quenched in a configuration short of minimizing the energy function. If Schelling's happiness is expressed as a function, this proposed energy function would be its negative, i.e., an unhappiness function. The low value of T would imply that only configurations that decrease the energy function (i.e., that increase happiness) will be accepted.

3.2 Schelling Model as a Derin and Elliott GRF

In general, the Potts energy function used by Derin and Elliott can be rewritten as:

$$\begin{aligned} U_{D\&E}(x) &= A + \sum_{c \in C} \zeta_c (n_{\diamond,c}(x) - n_{=,c}(x)) \\ &= A + \sum_{c \in C} \zeta_c (n_c - 2n_{=,c}(x)) \\ &= A + \sum_{c \in C} \zeta_c n_c - 2\sum_{c \in C} \zeta_c n_{=,c}(x) \end{aligned} \tag{7}$$

where A accounts for the energy contribution from single member cliques, which depends only on the individual proportions of sites occupied by each group and the clique parameters, α_k; C is the set of all cliques for this lattice L and neighborhood system, $\mathcal{N}$; each clique in C is of a particular type c (as in Fig. 2); ζ_c is the parameter for clique type c and the summation is over all clique types c in C; $n_{=,c}(x)$ and $n_{\diamond,c}(x)$ are the number of equal- and unequal-member cliques of type c, respectively, in the entire configuration x; and $n_c = n_{\diamond,c}(x) + n_{=,c}(x)$ is the total number of cliques of type c.

If the isotropic energy function used to produce the image in Fig. 4a is adopted, then the clique parameters simplify as in (8), where C_2 is the set of all two-member cliques.

$$\zeta_c = \begin{cases} \zeta > 0, & for\, c \in C_2 \\ 0, & otherwise \end{cases} \tag{8}$$

Since the number of cliques of each type c, n_c, is determined by the chosen model and not by any particular configuration of pixels in configuration x, so is the second term in final form of (7). If for the moment we assume that the histogram of site values is maintained throughout the MCMC process used here, so that A remains constant, then the first two terms can be replaced by a constant, say B, and (7) simplifies to (9), which will now be shown to be very similar to a reasonable unhappiness function for the Schelling process.

$$U_{D\&E}(x) = B - 2\zeta \sum_{c \in C_2} n_{=,c}(x) \tag{9}$$

Restricting our consideration to a Schelling board with no empty sites, and letting $n = |N_{ij}|$, the number of neighbors for any site, the process' happiness for a particular site x_{ij} in a configuration x is the number of neighboring sites with values equal to it. For a range one Moore neighborhood, a happiness function can be computed on the entire configuration as in (10), and a Schelling "unhappiness function" is simply its negative, as in (11).

$$\begin{aligned} H(x) &= \sum_{i,j} \frac{1}{n} \sum_{h,k \in N_{ij}} (x_{ij} = x_{hk}) \\ &= \frac{1}{n} \sum_{c \in C_2} n_{=,c}(x) \end{aligned} \tag{10}$$

$$U_S(x) = -\frac{1}{n} \sum_{c \in C_2} n_{=,c}(x) \tag{11}$$

Setting aside how switch candidates are selected, it is clear from (9) and (11) that the change in energy or unhappiness resulting from a site switch depends only on the change in the number of cliques with equal-valued members.

4 Texture Measures Applied to Segregation

The exposition above suggests a set of statistics that might prove useful in measuring the extent of segregation or clumping in a regular lattice. Here we describe such statistics for a 2D lattice L. These "clique organization parameters" [11], related to common measures of image texture called grey-level co-occurrence statistics [12], measure excessive organization in cliques. For each clique type, how many more or fewer equal member cliques are there, compared to what a random assignment of site values would yield? Extra equal valued cliques indicate segregation. (Extra unequal valued cliques would indicate a sort of "hyper-diversity." This is useful for describing images with alternating texture components. A social example might be the increased distrust of one's own group that occurs in highly diverse environments, described by Putnam [13].)

First assume the formation of display x as a random and independent assignment of values from $Q = \{q_1, \ldots, q_m\}$ onto L according to distribution $f(q_l)$. Then the probability $p_{2=}(x)$ that any two distinct sites, x_{ij} and x_{hk} in x have equal values is given by (12).

$$p_{2=}(x) = P(x_{ij} = x_{hk}) = \sum_{l=1}^{m} f(q_l)^2 \tag{12}$$

As above, assume a neighborhood system, $\mathcal{N}$, with C_2 the set of two-member cliques. Let n_2 be $|C_2|$, the number of pairs in the display, and $n_{2=}(x)$ be the number of these that are equal-valued in configuration x. Then $s_{2=}(x)$ as defined in (13) is a measure of the "super-equality" of pairs in x.

$$s_{2=}(x) = (n_{2=}(x)/n_2) - p_{2=} \tag{13}$$

Figure 5 presents three images with varying degrees of segregation. The first image, (a), is a totally random configuration of pixels from a uniform histogram; there is no organization in the image, and $s_{2,=}$ is almost 0. Images (b) and (c) are the results of running an MCMC method on an initial configuration like (a), using a Derin and Elliott energy function with parameters equal to 1.0 for all two-member cliques, and zero for all larger cliques. Such parameters lead to an isotropic pattern. Different cooling schedules were used to obtain (b) and (c): for (c) the temperature was lowered ten times more slowly than for (b), while for (b) the number of iterations at each temperature was ten times higher. The more rapid chilling quenches the pixels into the smaller regions found in (b) compared to (c)'s more homogeneous groupings. The statistic $s_{2,=}$ captures the higher level of segregation found in (c); it is 0.571 compared to 0.396 for (b).

The definition of $s_{2=}$ counts all two-member cliques; it does not consider their shapes and so it ignores direction. Excessive pairing in one direction could be counter-balanced by sparse pairing in another; defining measures for each type of two-member clique would assure that neither type of organization is lost. This could measure segregation along a geographic feature such as a north-south highway or a river.

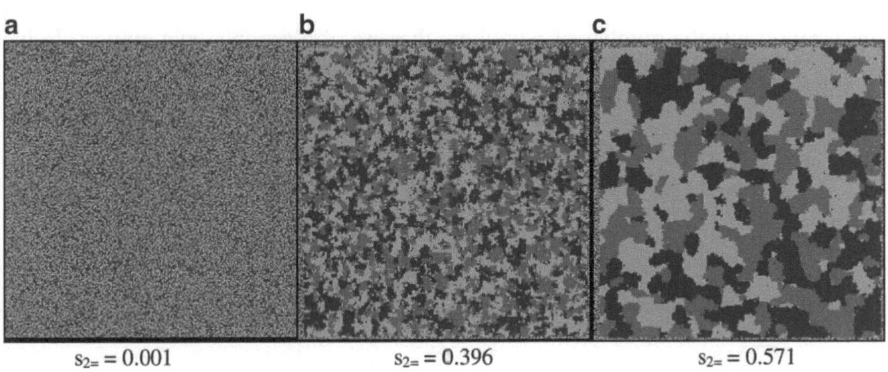

Fig. 5 An initial random image (**a**) and the results of simulating a highly probable configuration using Derin and Elliott's energy function [2], with different cooling schedules: for (**c**) the temperature was reduced 10 times more slowly than for (**b**); while for (**b**) the number of iterations at each temperatures was 10 times higher. The super-equality of pairs in each image, measured by $s_{2,=}$, captures the degree of segregation in the images

5 Application to the Output of an Agent Based Model

Preliminary tests support the idea that the image texture descriptor in (13) could describe clustering in patterns output from agent based models. Wetlands [4] is an ABM that investigates how group memory affects sociality and collective intentions. It is implemented in MASON (Multi-Agent Simulator of Networks and Neighborhoods), created at George Mason University [8]. In Wetlands, a landscape endowed with a simple weather system provides food, moisture, and shelter from rain. Agents are culturally homogeneous groups that roam this landscape of hexagonal sites, searching for provisions and shelter. They are from two different cultures, and they share information about supportive sites with other groups of their own culture. This information sharing within cultures should lead to clustering by culture as members head toward the same beneficial sites.

Figure 6 shows the output from a Wetlands run that had 500 groups, half from each culture, represented by the colors grey and black. The landscape, moisture and food layers are turned off to highlight the positions of the groups. This run was on a 50 by 50 toroidal grid of sites, 78 of which were inaccessible since they represented bodies of water, leaving 2,422 accessible sites. Run parameters were set for a range one hex neighborhood as depicted in Fig. 1.

In applying (12), we consider the empty sites as a separate possible site value (white), so $Q=\{G, B, W\}$. There are 250 grey sites, 250 black sites, and 1,922 empty, or white, sites. These led to $p_{2=}=0.651$, the probability that any pair of sites is equal-valued, if values were just uniformly randomly assigned to the sites.

To get n_2, the total number of two-member cliques, subtract the 78 inaccessible sites from the total 2,500 sites and multiply by 3, the number of two-member clique types for a range 1 hex neighborhood system (see Fig. 2), obtaining 7,266. (This is a slight over count, since it allows cliques that straddle one side of the small inaccessible regions; this was considered insignificant.)

Wetlands produces a number of statistics as the simulation steps along. One of these represents the number of neighbors that are of a group's own culture. Using

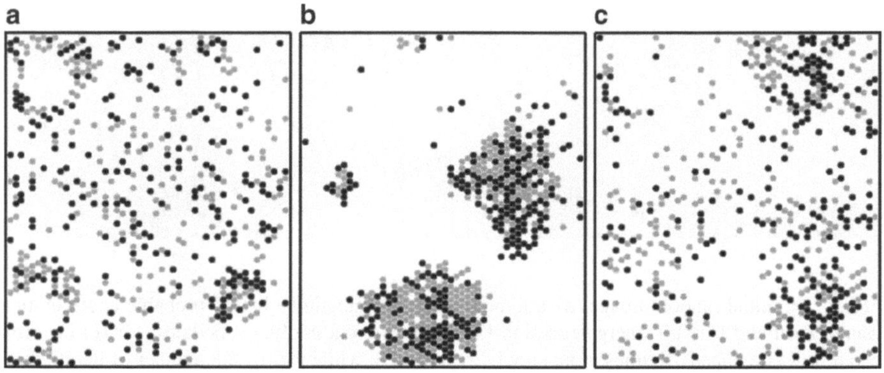

Fig. 6 Output patterns of groups from two cultures from Wetlands simulation

this value and the total number of groups, we can compute $n_{2=}(x)$, the number of equal two-member cliques for the output pattern x from any step of the simulation. This enables us to compute the excessive equality of pairs in a pattern, according to (13). For output patterns (a), (b) and (c) in Fig. 6, rather than $p_{2=}=0.651$ (the proportion of equal two-member cliques we would expect if there were no clustering by culture), the proportions were 0.825, 0.878, and 0.830, respectively. This shows that the groups in the Wetlands simulation are clustering by culture, with an excess of like-culture pairs of $s_{2=}=0.174$, $s_{2=}=0.227$ and $s_{2=}=0.179$, respectively. It is also interesting to note that output pattern (b), which appears the most tightly clustered, has the highest excess of the three patterns, 27% higher than (a) and 30% higher than (c).

6 Applications to Social Simulation Beyond the 2D Grid

This paper compares the Schelling segregation process, as an example of an agent-based model on a 2D grid, to the simulation of an image texture, as an instantiation of a Gibbs Random Field. It describes applying measures useful for the latter to the former and to another agent-based model, Wetlands. This comparison is a subset of those we presented earlier [14], which connected each of these two processes with two others that are highly relevant to social simulation. That presentation showed how closely these two processes are paralleled by adaptive learning in potential games with spatial interaction, studied in evolutionary game theory. These three models in turn are all connected, by way of the Gibbs distribution, to one of the most theoretical sides of social network analysis (SNA). Table 1 summarizes the comparisons made there among the four processes.

For many years, SNA researchers have applied the Hammersley-Clifford theorem to support using the Gibbs distribution, known to them as p* ("p-star"), to describe their networks [15]. Neighbors are defined by a set of edges representing social relationships, and as in the GRF, cliques are sets of mutual neighbors. They now consider p* to be one of a family of distributions called exponential random graphs models (ERGMs). Recently they have been developing advanced MCMC estimation methods for the terms of their models [16, 17]. This newer approach, while difficult, is proving to be very fruitful, enabling statistical inference and the avoidance of degenerate graph models. The clique organization parameters discussed here are simple versions of such ERGM terms.

7 Conclusion

We have presented measures, clique organization parameters, used to model texture in images, and have shown how they can express the degree of clustering in the output of an agent-based model, Wetlands. The connection between the uses for

Table 1 Comparison of the Schelling model and image synthesis to other important social simulation processes

	Schelling model	MRF/GD image synthesis	Evolutionary game theory	Exponential random graphs for SNA
Model	1D, 2D lattice, full or empty sites	2D lattice of sites, neighborhood system	Potential game played on a graph	General graph defining neighbors
Process	Sequence of position exchanges	MCMC synthesis	Sequence of plays based on histories or nbhd sampling	MCMC parameter estimation
Driving Function	Happiness	Energy sum of clique potentials	Utility based potential	Energy
Noise Presence	None	Increasing energy allowed	Not always best reply (trembling hand)	Increasing energy possible
Benefits to Share	Social Relevance	Visualization, cooling schedule experiments	Most general? Social and strategic interpretation	Advances in parameter estimation

these measures derives from the similarity of the underlying processes, which are based on local attractions. In fact, more varied patterns that also involve local repulsions can be represented when a full set of cliques is considered [11]. The measures discussed here are a very simple version of the terms estimated by social network modelers, but the analogy with their ERGMs supports using the measures to describe clustering patterns on 2D grids. Future efforts will involve performing more experiments to further test the value of these measures.

Acknowledgments The author thanks Prof. Rob Axtell and Prof. Claudio Cioffi-Revilla for their kind encouragement to pursue these ideas; Gabriel Balan for his help with using Wetlands; Andrew Pérez-López for a refresher course in Java; and two sets of anonymous reviewers for thoughtful, helpful comments.

References

1. Geman S, Geman D (1984) Stochastic relaxation, Gibbs distributions, and the Bayesian restoration of images. IEEE Trans Pattern Anal Mach Intell 6:721–741
2. Derin H, Elliott H (1987) Modeling and segmentation of noisy and textured images using Gibbs random fields. IEEE Trans Pattern Anal Mach Intell 9:39–55
3. Schelling TC (1971) Dynamic models of segregation. J Math Sociol 1:143–186
4. Cioffi-Revilla C, Paus S, Luke S, Olds JL, Thomas J (2004) Mnemonic structure and sociality: a computational agent-based model. In: Conference on collective intentionality IV, Certosa di Pontignano, Siena, Italy, 13–15 October 2004
5. Langseth H, The Hammersley-Clifford theorem and its Impact on Modern Statistics. www.idi.ntnu.no/~helgel/thesis/forelesning.pdf
6. Feller W (1968) An introduction to probability theory and its applications. Wiley, New York
7. Van Laarhaven PJM, Aarts EHL (1989) Simulated annealing: theory and applications. Kluwer Academic, The Netherlands
8. Luke S, Catalin Balan GC, Sullivan K, Panait L, MASON. http://www.cs.gmu.edu/~eclab/projects/mason/. Accessed 14 July 2009
9. Luke S, Simple Schelling segregation. http://cs.gmu.edu/~eclab/projects/mason/projects/schelling/. Accessed 14 July 2009
10. Schelling TC (1978) Micromotives and macrobehavior. W.W. Norton & Co., Inc, New York
11. Pérez-López K, Sood A, Manohar M (1994) Selecting image subbands for browsing scientific image databases. In: Proceedings of the conference on wavelet applications, part of SPIE international symposium on OE/aerospace sensing, Orlando, FL, 5–8 April 1994
12. Haralick RM, Shanmugam K, Dinstein I (1973) Texture features for image classification. IEEE Trans Syst Man Cybernet 3(6):610–621
13. Putnam RD (2007) E Pluribus Unum: diversity and community in the twenty-first century. The 2006 Johan Skytte Prize Lecture, Nordic Political Science Association
14. Pérez-López K (2007) The texture of a network: adapting image texture measures to describe Schelling segregation on complex networks. Presented at the international conference on economic science with heterogeneous interacting agents, Fairfax, VA, 17–20 June 2007
15. Frank O, Strauss D (1986) Markov graphs. J Am Stat Assoc 81:832–842
16. Wasserman S, Robins G (2006) An introduction to random graphs, dependence graphs, and p*. In: Carrington PJ, Scott J, Wasserman S (eds) Models and methods in social network analysis. Cambridge University Press, Cambridge, pp 148–161
17. Snijders TAB (2002) Markov chain Monte Carlo estimation of exponential random graph models. J Soc Struct 3(2):1–40

References

Autonomous Tags: Language as Generative of Culture

Deborah Vakas Duong

Abstract The first intelligent agent social model, created by the author in 1991, used tags with emergent meaning to simulate the emergence of institutions based on the principles of interpretive social science. This symbolic interactionist simulation program existed before Holland's Echo, however, Echo and subsequent programs with tags failed to preserve the autonomy of perception of the agents that displayed and read tags, as first program did. These subsequent tag programs include Epstein and Axtell; Axelrod; Hales; Hales and Edmonds; Riolo, Cohen and Axelrod, as well as the works on contagion originating in Carley, etc. The only exceptions are the author's 1995 SISTER program, and Axtell, Epstein, and Young's 2001 program on the emergence of social classes, which was influenced by the symbolic interactionist simulation program at George Mason University, and Steels' 1996 work. Axtell Epstein and Young's program has since been credited for strong emergence (Desalles et al.). This paper explains that autonomy of perception is the essential difference in the symbolic interactionist implementation of tags that enables a strong emergence to occur, and that is why strong emergence has occurred in the works of Duong and of Axtell, Epstein and Young. This paper explains the important differences in existing tag models, pointing out the qualities that enable symbolic interactionist models to become social engines with strong emergence, and also introduces new work that puts the SISTER program in a spatial grid, and explores what happens to prices across the grid. In half of the runs, a standard of trade, or money, emerges.

Keywords Symbolic interactionism • Interpretive social science • Emergent language • Agent-based simulation • Tags • Autonomy

---

D.V.D uong(✉)
Science Applications International Corporation, 1525 Wilson Blvd., Suite 800, Arlington,V A2220 9U SA
e-mail:de borah.v.duong@saic.com

K. Takadama et al. (eds.), *Simulating Interacting Agents and Social Phenomena: The Second World Congress*, Agent-Based Social Systems 7,
DOI 10.1007/978-4-431-99781-8_17, © Springer 2010

"In the beginning was the Word, and the Word was with God, and the Word was God. The same was in the beginning with God. All things were made by him; and without him was not any thing made that was made. In him was life; and the life was the light of men. And the light shineth in darkness; and the darkness comprehended it not." (John 1:1–5, KJV)

1 Introduction

Holland saw the creative power of the word as important in the formation of living systems when he included the tag as one of the three basic mechanisms of complex adaptive systems. A "tag" is simply a sign, such as a name or a physical trait, which is used to classify an agent. In the social world, a tag may be a social marker, such as skin color, or simply the name of a social group. A tag goes hand in hand with the other two mechanisms Holland thought important to complex adaptive systems: an internal model (whether tacit or explicit) to give meaning to tags, and building blocks to accumulate and recombine the structures that result from those meanings into hierarchical aggregates [18]. Holland's insight on the generative nature of signs was correct, however, his implementation of the sign, in his version of tags, failed to include perceptual autonomy, and as a result, did not have the generative power of signs. The Symbolic Interactionist Simulation program demonstrates that it is necessary to distinguish between the sign and the interpretation of the sign to create a full social engine. In accordance with interpretive social science, the keys to the engine lie in interpretation.

Holland is commonly thought to be the first to use tags to simulate social phenomena. However, there is another variation on tags, the symbolic interactionist simulation technique, that was developed before Holland's complex adaptive system research program, the Echo project [6,19]. Like Echo, symbolic interactionist simulation recognizes the primacy of signs in the formation of living systems, but differs from Echo in that its agents have autonomous perception of the meaning of signs. The difference is understandable, because the principle of autonomy of perception is more prominent from the social sciences standpoint than from the biological standpoint, even if it exists in biology as well [22]. Many of the ideas in microsociology are inherited from phenomenology and hermeneutics, philosophies that contemplate the mysteries of autonomy, such as the paradox that human beings can only interpret meanings through their individual experiences with their senses, and yet they still come to share meaning [27]. This "hermeneutic paradox" is the core issue of micro- macro integration in sociology from the angle of perception: to solve the hermeneutic paradox is to solve the mystery of the "invisible hand" by which autonomous, selfish agents synchronize their actions into institutions for the good of the whole. Since emergence in agent-based social simulation is fundamentally about solving the micro macro link, symbolic interactionist simulation seeks to solve the hermeneutic paradox. It is by virtue of the preservation of autonomy that symbolic interactionist simulations exhibit strong emergence and constitute minimal social engines.

In Holland's Echo program and its successors that simulate the emergence of cooperation in iterated prisoner's dilemma (IPD) programs, tags are implemented with replicator dynamics. Referring to the work of Riolo, Cohen, and Axlerod as well as the work of Hales and Edmonds, Hales discusses the tag implementation: "the models implement evolutionary systems with assumptions along the lines of replicator dynamics (i.e. reproduction into the next generation proportional to utility in the current generation and no 'genetic style' crossover operations but low probability mutations on tags and strategies)" [16]. Replicator dynamics do not keep the principle of autonomy of perception: one agent interprets a sign the same way as another agent because they have a common ancestor, not because they both induced the sign separately based on their individual experiences. Simulations of the emergence of common meaning of tags using replicator dynamics exhibit high amounts of genetic linkage (biological or mimetic), so that the relation between the sign and the behavior is an artifact of the method, rather than emergent from the simulation.

Any simulation of contagion that explains macro level institutions with micro-level imitation does not exhibit strong emergence by definition. If we use a general definition of institutions as "Systems of established and prevalent social rules that structure social interactions" [17]. Then a model of simple contagion would explain the prevalence of institutions as an aggregate of copying behavior rather than as emergent phenomena which is more than the sum of its parts. Simulations of contagion, such as in the works of Carley, may involve homophilia that emerge fractious structures, however, these structures are faults in cohesion, explaining where institutional cohesiveness breaks down rather than explaining why social rules are prevalent. To the extent that they are prevalent in contagion/replicator dynamic simulations is the extent to which the are aggregate as opposed to emergent phenomena. Micro macro integration sociologist James Coleman challenges us to go farther than cohesion based theories of institutions, to theories based on (broadly defined) structural equivalence. In fact, cohesion vs. structural equivalence is a controversial divide in the theories of social homogeneity (Friedkin). Coleman believes that to explain institutions, we must explain the arise of a network of relations in a social system, that is, how institutions are different depending on roles, which together constitute a social system and not just an aggregate [4]. For example, Coleman claims Weber's theory of capitalism as caused by the protestant work ethic is not a theory of emergence because it is an explanation of the propensities towards the same behaviors, rather than an explanation of how the different behaviors in the capitalist system came be. In other words, to explain Macro phenomena from Micro, we must explain how an entire system comes to form, and only this is "strong emergence". In simulations with contagion, copying is one of the assumptions, and so is not explained by the simulation. In order to be emergent, phenomena cannot be derived directly from the assumptions and starting conditions of the simulation, but must be of a higher order.

Bedau [2] and other philosophers of emergence agree that "emergent properties have irreducible causal power on underlying entities "Downward causation," or

"Immergence" as Gilbert called it, is the result of the whole being more than the sum of the parts: the lower parts of the hierarchy are restrained and act in conformity to the laws of the higher level. The influence of the macro on the micro is irreducible in that it is directly downward as opposed to influencing via the micro properties. Thus, two way causation, upward because the macro is composed of the micro, and downward because there are upper level laws that limit the micro, is necessary for emergence in the strong sense. Desalles, Galam and Phan [5] believe that for downward causation to occur in the case of social emergence, agents must be equipped to identify emergent phenomena before it can affect their behavior, and Muller adds autonomy of perception: that this must be through the physical world (of physical signs coming to have new meanings), rather than by direct copying of other agent's perceptions [23], so that agents may be free to interpret the world to achieve their individual goals. According to Desalles et al., agents must describe the emergent phenomena they observe in a language other than the language of the lower level process itself, and agents must have a change of behavior that feeds back to the level of observation of the process. This insightful definition of strong emergence acknowledges the importance of autonomy of perception, that is, of not allowing agents to copy each other's internal states, in developing a new emergent language to describe emergent phenomena, that is different than the language of the lower level phenomena. To use the same language as the micro would be to use the same objects, relations, and patterns as the micro, but emergent phenomena requires an entirely new set of objects, relations and patterns, with new concepts to describe them. In agent-based systems, the first rudimentary language to describe un-preprogrammed, emergent phenomena, is the tag that comes to mean the macro level phenomena that emerges during the simulation. As institutions emerge, so should tags emerge to represent them, and to influence micro level behaviors. Immergence, or the ability of the lower level agent to change its behavior based on the emergent social phenomena, opens the door for generative feedback between micro and macro social levels. Such a generative engine, which some social scientists would call a dialectic, characterizes strong emergence.

Luc Steels' research program also addresses the hermeneutic paradox: his agent's signs come to have shared meaning, even though they have autonomous perception. However, his agent's signs were not tags related to social structure as in symbolic interactionist simulation. In Steel's work, arbitrary signs come to have meaning as agents use them to differentiate objects by their features. As individuals make distinctions based on their own perceptions and associations, they come to have shared words to refer to features and shared ontologies of what distinctions to make are important, in an emergence with upper lower feedback [26]. Ironically, even though these agents may be embodied as robots, they are not truly situated, as they are describing their environment but not applying this description to their utility, or in anyway changing their world with their language. The ontologies these agents use to cut up the world are arbitrary, whereas the ontologies of human languages cut up the world based on utility. Although language is reproduced, culture and the way that the world is manipulated is not.

2 Symbolic Interactionist Simulation

In symbolic interactionist simulation, the mechanism of autonomous emergence of the meaning of signs (tags) facilitates a strong emergence of practical ontologies that coevolve with practical behaviors. Symbolic interactionist agents interpret signs based on utility, so that an interpretation makes sense given the background of the agent's individual experiences. In symbolic interactionist simulation, a sign is interpreted in a certain way because it makes utilitarian sense, and not because it is copied. Agents communicate solely through signs, inducing the meanings of both displayed and read signs. Inductions are based on economic and practical gain, and as a result of these utilitarian interpretations, symbol system and social institutions coevolve. The interpretation of the sign is more important than the sign itself because the individual, autonomous, interpretation causes cultural reproduction.

The first symbolic interactionist simulation [6,7] was a simulation of a workforce of employers and employees. In some of the runs, for example, there were three employers and 50 employees in a society. Each employee had either a high or low level of talent, which the employer could not see until after the employee was hired. However, the employer would look at the signs that an employee displayed to guess whether that it was talented. The prediction was based on the employer's individual past experiences with employees. The employee displayed a fixed sign (such as skin color, race or gender), a sign that costs money (such as a new suit) and a sign that is free (such as a fad). The fixed sign was made to be uncorrelated with talent. Employees obtained money through employment, and thus employees that could stay employed longer could make more money than employees that were fired frequently. A certain percentage of the workforce of each employer was laid off every cycle, but employees that were not talented were laid off in greater proportions. Thus, an employee that is talented has more of a capacity to make money, and the potential to differentiate itself from a non-talented employee using that money. The employees would choose a set of signs to display based on their prediction of whether they would be hired after an employer saw them. This prediction was based on their individual past interviews outcomes. Of course, employees could only display the purchasable signs that they could afford. Both the employer and the employee agents had IAC neural networks to induce the meanings of the signs based on their private experiences with the signs. Even though the signs were arbitrary and autonomously perceived (employers did not consult each other on the meanings of signs, nor did employees), they came to have a shared meaning. Agents learned to buy expensive suits as status symbols, and race often became an issue despite the fact that race was uncorrelated with talent, because it sometimes became correlated with the suit. Races could get into a vicious circle where they could not afford a suit because they were not hired and were not hired because they did not wear a suit, at which time social classes based on race would form.

In this symbolic interactionist simulation, interpretation is generative of culture. Because individual employers are free to interpret the meanings of signs on the basis of individual experiences, individual interpretations may differ. An employer's discovery of talent in a race which he formerly believed to lack talent allows the talented members to afford suits, raising their esteem in the eyes of other employers. Individual interpretation allows self fulfilling prophecy to take effect as social change. Autonomous interpretation of tags is the means by which downward causation takes effect, the means by which upper level patterns change lower level behaviors based on individual utility.

Axtell, Epstein and Young's model of the emergence of social classes subsequently adapted the autonomy of the symbolic interactionist tag methodology [1]. Desalles et al. [5] took note of the strong emergence Axtell et al. achieved by use of autonomously interpreted tags. Axtell et al. achieved the emergence of social class based on fixed tags (such as skin color, race or gender) in a one shot bargaining model.

3 SISTER

Another symbolic interactionist simulation which uses a one shot bargaining model, SISTER (Symbolic Interactionist Simulation of Trade and Emergent Roles) was prior to and influential on Axtell, Epstein and Young's work on the emergence of social classes [14]. SISTER also simulates the coevolution of symbol systems and social structure [8–12]. The role of interpretation in the reproduction of culture is even more important than in the 1991 Symbolic Interactionist model, as the self fulfilling prophecy of interpretation in SISTER causes roles and specific role knowledge to be reproduced. Diversity of interpretation through autonomous perception is vital to a distributed definition of roles and to the reproduction of cultural knowledge such as how to cook.

SISTER is a simulation not of a modern economy, but of a simple barter economy, where no wealth may accumulate from day to day. Agents produce in the morning, trade in the afternoon, and consume at night, leaving nothing for the next day. It is assumed that agents seek to have eight goods in equal amounts, and as much as they can get of them. Agents can produce some or all of the goods as they please, but these activities cost effort. An agent has only limited efforts to spend, but by concentrating efforts on a small number of categories of goods then more total units of good will be made. This simulates the benefits of specialization, also known as economies of scale. By this design the agents will be happier if they make lots of one good and trade it away for the others, however, it is up to the agents to learn how to trade. They develop institutions in the process of learning how to trade, and their institutions are solutions to the problem of how to trade. They start out the simulation completely ignorant of what to produce, what to trade, how much to trade, and whom to trade it with, and what sign they should present to others to tell who they are. The knowledge they come to have to get to the answer, the development of interlocked behaviors and shared meanings prerequisite to that answer, are the emergent institutions. This study focuses on just one of these institutions: the emergence of a division of labor. Other institutions that the agents develop are price and money.

The subjective micro-level rules come from the principles of interpretive social science as found in phenomenological and symbolic interactionist sociology. The agents follow a basic principle of interpretive social science: they are closed with respect to meaning. That is, they cannot read the minds of other agents to learn how to interpret signs, but can only understand signs through associations with sensory phenomena that they experience in their individual lives. They each have their own private inductive mechanism with which they perform the symbolic interactionist task of inducing what signs they should display and the meaning of signs that they read. Their inductive mechanisms are autonomous in the sense that they are not seeded with information from other inductive mechanisms. With these inductive mechanisms, the agents interpret the signs they display solely from the context of their individual previous experiences, never copying another agent's sign, or another agent's interpretation of a sign. Yet despite this autonomy, the signs emerge shared meaning. This is in accordance with the hermeneutic paradox of our inventing our own meaning and yet sharing meaning. As in the symbolic interactionist paradigm, the signs come to have meaning as is pragmatic for the agents in the function of their "everyday life". The signs they learn to read and display are signs of whom to approach to make a trade with and what to display to attract trade. The meanings that the signs come to have are roles in a division of labor. This is in accordance with phenomenological sociologist Peter Burger's emphasis on the importance of roles and symbols of roles as a mechanism of organizing behavior. To learn how to trade, every agent must learn his role in terms of what goods to produce or assemble, and have general knowledge of the other roles in the society so that he may know who to trade with.

3.1 The Design of SISTER

SISTER is a simulation of daily habits of trade in a bartering society. Every agent learns a plan of harvesting, cooking, trade, and a sign to display for a single day of the simulation. Each agent has a limited number of "efforts" that it learns to allocate between harvesting, cooking and trading of goods. It allocates these efforts to specific goods to harvest or cook, and specific trade plans to perform. At the beginning of the day, agents harvest according to their production plans. They may harvest a few of each good, or more of some and less of others as their plans dictate.

Next, agents trade. All agents have both active and passive trade plans that are activated if agents devote the right amount of effort to them. Each trader follows its plans for active trading, by seeking out agents to trade with that display signs closest to the ones in their trade plans. If the passive trader has a corresponding trade plan and both the active and passive trader have the goods, then the trade takes place. See Fig. 1 for an illustration of corresponding trade plans. If the scenario simulates composite (cooked) goods, then an agent must have all the components of a good before it can trade that good [20]. An agent may trade to get the ingredients of a recipe, cook it, and trade it away.

Fig. 1 Corresponding Trade Plans. Agents trade with agents who have corresponding trade plans and are wearing the correct sign

In the evening, the agents consume all of their goods, and judge their trade plans for the day solely on their satisfaction with what they consume that night.

The agents' motivation for trade lies in the encoded microeconomic rules. This simulation encodes two concepts from micro economics: the nature of the agent efforts and the nature of agent needs. The nature of agent efforts is that if agents concentrate their efforts on fewer activities, they are able to be more productive than if they spread their efforts out over more activities. This simulates the benefits of specialization, or economies of scale. The nature of agent needs is that agents seek an even spread of several goods, and as much as they can get of them. It is as though each of the goods is one of the four essential food groups, and they seek a balanced diet of those goods.

Economies of scale are encoded by setting the efforts devoted to an activity to an exponent, whether that activity is a particular trade, the harvesting of a good, or the combining of goods (cooking):

$$activity = Kt^{c}$$

The number of specific activities an agent may complete is equal to a constant for that type of activity, K, times the efforts designated to that particular activity, e, raised to an effort concentration factor for that type of activity, c. For example, if the trade constant is 0.5 and the trade concentration factor is 3, and an agent devotes 2 efforts to trade four units of oats for three units of peas, then there are 0.5(23) = 4 such trades it can make. If the activity is a trade, the concentration of effort might mean that the agent has invested in a cart to make an active trade, or in storage to make a passive trade. If the activity is harvesting, it might mean that the agent has put in the investment to harvest a particular good well. If the effort is composing goods (cooking), the agent may have invested in training to learn how to cook a particular type of food. Whatever activity it is, this equation means that putting a little more effort into the activity will make it much more productive.

The nature of an agent's desires are encoded with the Cobb-Douglas utility function. At the end of the day, each agent consumes all of its goods. How happy an agent is with its trade plans is judged solely by the Cobb-Douglas utility function of the goods an agent consumes:

$$utility = \prod_{i=0}^{n} good_i^{weight_i}, \quad \text{where} \sum_{i=0}^{n} weight_i = 1.0$$

"Good" is the amount of a good that an agent consumes, n represents the number of different goods, and weight is a measure of how much each individual good is desired. All of the weights add up to one. Each agent has a minimum of one of each good given to it. This is a standard utility function in economics. If all of the weights are the same, it makes it so that agents want a spread of all different types of goods, and as much as they can get of them. The agents want a spread of goods in the same sense that people need some of each of the four food groups. For example, if an agent has eight units each of two of the four food groups, his happiness is $80.25 \times 80.25 \times 10.25 \times 10.25 = 2.82$. If the goods are more spread among the food groups, and the agent has four units of each of the four food groups, then its happiness is $40.25 \times 40.25 \times 40.25 \times 40.25 = 3.48$. The agent would rather have four of four goods than eight of two goods. With this equation both the spread of goods and the amount goods are important. In this study, the weight for each good is equal, so that differences in outcome cannot be attributed to uneven utility values for individual goods.

The equations for effort and utility make it to the agents advantage to concentrate their efforts on making fewer goods so that they can make more of them, and then trading them for the rest of the goods, so that they can have an even amount of many goods. Agents may choose to make all of the goods for themselves until they learn how to trade them. In order to learn to trade the goods, the agents must learn to read each others signs and to display the correct sign. This is done according to the rules of interpretive social science. Agents never copy another agent's interpretation of a sign, but rather interpret and display signs solely according to its own utility. To simulate Parsons' double contingency, the sign is double induced: both the sign to seek for a trade and the sign to display on ones self are learned [24]. These signs come to have common meaning as the agents differentiate themselves. As in Parsons' theory, the ordering of society comes through a shared symbol system, and as in Berger's and Coleman's theory, that ordering is a system of roles [3,4,24].

The passive trader displays a sign that its individual genetic algorithm has learned. The sign that an agent displays for its passive trades comes to represent that agent's role. That sign starts out random, but comes to signify a role when all the agents that have learned to display the same sign have similar behaviors. Figure 2 illustrates agents that have differentiated into roles denoted by a learned tag that they display in common. An endogenous differentiation of the agents occurs. This happens because the genetic algorithms are individual to each agent, and not seeded with information from other agents, in contrast to tag programs like Echo in which a single genetic algorithm is used for all agents. Rather, the SISTER genetic algorithms coevolve. Several individuals become replacements for each other in their behavior. An agent who wants to make an active trade can teach (that is, exert "selective pressure" on) many different agents who can replace each other to make a trade.

For example, let's call the goods oats, peas, beans, and barley. Suppose an agent in the simulation has produced more oats and fewer beans. Suppose also that he displays a random sign to attract trade, and another agent with more beans and fewer oats guesses the meaning of that sign by coincidence while trying to trade his beans for oats. Both agents are satisfied with the trade and the sign: they remember this, and repeat it. The more the trade is repeated, the more it becomes a stable thing in the

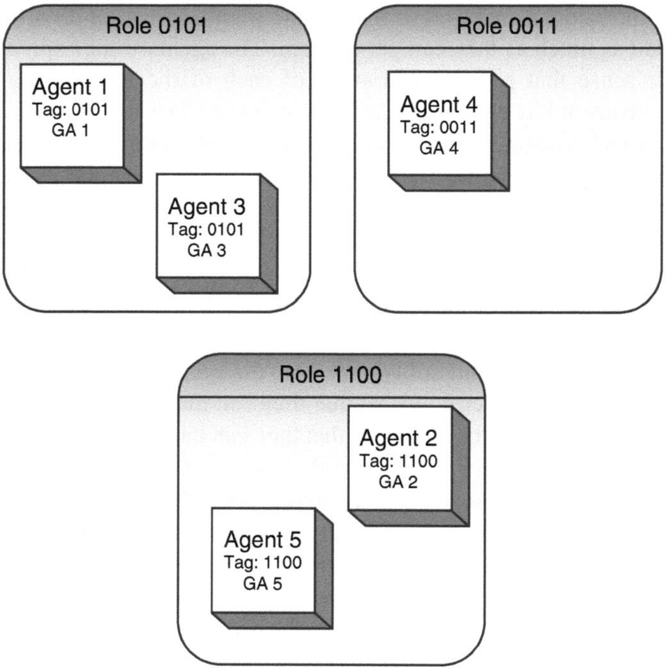

Fig. 2 Agents differentiate into roles. Roles are designated by tags, learned from an agent's individual GA. Different agents which have the same tag are said to be members the same role if the agents who display the same tag also have the same behaviors. These tags are individually learned by each GA, but come to mean the same set of behaviors. The individual agents have autonomy of perception in that they learn the meaning of the tags with individual genetic algorithms that are not seeded from other agents

environment for other agents to learn. Since an agent with an active trade plan is looking for any agent who displays a particular sign, then any agent can get in on the trade just by displaying the sign. The agents come to believe that the sign means "oats," in the sense that if an agent displays the sign accidentally, the other agents will ask him to sell oats. This will put selective pressure on that agent to make and sell oats. At the same time, other agents who produce oats will benefit from learning the oat sign so as to attract trade. After a while, the society divides into roles, with groups of agents displaying the same sign and having the same behavior. If a new agent comes into the simulation, then to participate in trade he must learn the sign system already present. The signs are a guide to his behavior: when he displays a sign, the other agents pressure him to have the corresponding behavior. Thus a sign creates expectations of behavior, in accordance with Luhmann's model of expectation expectations. A system of signs represents a compromise between the interests of the agents who have put selective pressure on each other to behave in ways beneficial to themselves. These signs enforce evolutionary stable strategies of agent behavior.

If composite goods or "cooking" is in the scenario, then agents must have all of the goods that a composite good is made of before it can trade that good. In a scenario with composite goods, agents have to know more to make trades in that good than they have to know in simple scenarios. They have to know to either harvest or purchase the goods that a composite good is made of, in addition to having to know who would buy the composite good. If a newly inserted agent displays the sign of one who sells a composite good, then it learns the component parts of the good when other agents come to it to sell those components. An agent is thus trained in how to perform his roles by the expectations of the agents in roles that interact with that role. For example, suppose that in a SISTER society, the harvested goods were corn and lima beans, and the composite good was succotash, made from corn and lima beans. Suppose the agent who discovers succotash puts up a sign that she sells succotash. She buys lima beans and corn from lima bean and corn farmers, and finds that she has much business when she sells her succotash to the local diner, that uses her succotash to compose some of their dinner entrées. Through experience, our inventor of succotash has learned who sells the components of her good, lima beans and corn, as well as who buys the components of her goods, local diners. New agents who want to sell succotash now do not have to relearn all of that, because as soon as they put up a sign that says they sell succotash, lima bean and corn marketers start to call them. The new agent, because of the other agent's expectations, figures out how to make succotash if he didn't know before. If he only knows about the lima beans when he starts to display his sign, he will quickly learn about the corn. This is because he will feel selective pressure to buy corn. It will make the succotash better as far as the diners are concerned, and will give him more business. And it will be easy to buy because corn agents are constantly asking him to buy it. If it comes to be to his advantage, he will learn it. This is how the knowledge of how to make composite goods is held in the mutual expectations of agents. The mutual expectations that the agents have of the roles allows individuals to take advantage of what other individuals have learned in previous interactions. The knowledge of the society is held in the mutual expectations of the symbol system, as in Parsons' and Luhmann's theories [21,24].

The reason that role systems can hold more information about how to make cultural products is that agents can replace one another and can learn from the expectations that other agents have of their replacement class. This is how they become trained to do their role. However, this training is not all encompassing: what they do is ultimately connected to their utility. They can reject trades if it is not to their advantage, for example, if they find that succotash tastes better with tomatoes than with lima beans.

Thus, knowledge of how to make cultural products is reproduced through a variety of points of view, a variety interpretations of signs which ultimately come from individual utility perceived autonomously. These interpretations at first occur on an individual basis, but then become coercive because of the strength in numbers of interpretations. Signs come to have meaning only as new social institutions begin to form, and the individual interpretation of signs generates these institutions. Autonomy of perception preserves the different points of view needed to hold all of the role based knowledge, and preserves the connection to utility.

3.2 *Mutual Information to Measure Roles*

To measure the amount of knowledge held in a society, the concept of mutual information, from information theory, is used. Mutual information is a measure of the information contained in a sign. If there is only one sign, and the agents that display it have several different behaviors, that sign indicates nothing. Conversely, if all sorts of signs all mean the same behavior, then those signs indicate nothing. If p(x) is the probability (frequency) of occurrence of a sign and p(y) is the probability (frequency) of a behavior, then mutual information is a measure of the correspondence of symbols to behaviors in a system [24]:

$$MutualInformation_{x,y} = \sum_{x}\sum_{y} p(x,y) \times \log_2 \frac{p(x,y)}{p(x)p(y)}$$

If there are many signs that mean many different behaviors, then the mutual information of a system is high. For example, if every agent that sells apples displays sign 1 and every agent that sells pears displays sign 2, then there is more information than if all agents display sign 1 and sell both pears and apples, or if some agents that sell apples display sign 1 and other agents that sell apples display sign 2. Mutual information is shown to correlate with utility.

3.3 *Experiment*

Hypothesis: When an agent society is distributed geographically such that trade over distance costs more, knowledge of how to trade complex goods is spread more easily when agents have role recognition than when they have individual recognition.

This experiment compares agents that use tags for recognition of roles, with agents that are only identified as individuals, over a geographical distribution. Duong and Grefenstette [11,12] showed that role recognition increased agent utility over individual recognition in non-spatially based trade, and Duong [13] showed that roles helped carry on culture over time despite the deaths of individuals. These works show that, if agents are not recognized as individuals, but as a member of a role by agents seeking other agents to trade with, they learn networks of trade better than if they only recognized each other as individuals. The difference between individual recognition and role recognition is the difference in going to "Joe" to trade some beans for oats opposed to going to "an oat farmer" to trade beans for oats. In this study, it is determined that roles help maintain cultural knowledge over geographical distance.

In all experiments, average utility is used to assess how well a society of agents trades. In SISTER, this is the same as average fitness. A t-test is done on the average utility over a run, as well as for every corresponding cycle in a run. This t-test compares the role recognition treatment to the individual recognition treatment. The t-test measures how unlikely it is that the individual recognition treatment is as good as the role recognition treatment given all of the data from the 20 experiments. If the probability is below 5%, then we have significant evidence that agents in societies that use role recognition generally have higher utilities than agents in societies that use individual recognition.

Another test done in all the experiments is a t-test for difference in mutual information between the treatments, to check if the increase in utility is due to an increase in mutual information. A correlation between mutual information and utility is taken to support the hypothesis that the reason behind improved trade is the better information of the role treatments.

This experiment will be like [13], except the complexity added will be a geographically based cost of trade, instead of death of agents. Agents looking for active trade will have to pay more for trade over distance. Efforts must be devoted for a trade in the simple scenarios (one unit of effort for each trade) but in this experiment, the number of units of effort required increases with Manhattan distance between the agents in a 4 by 4 grid. In spatial scenarios, a store four blocks away costs four times as much to reach as a store one block away. Agents can only travel to make a trade as far as they have efforts devoted to making that trade. For simplicity, the goods that an agent sells as a passive trader remain in the same location throughout the simulation. An agent has a "store," through which it can sell goods in a single stable place, even though it travels to make purchases. An agents "store" remains in the same location for passive trading, but the agent himself must travel to other stores for active trading. For example, a tomato farmer may have set up shop on fifth street, and can still sell tomatoes passively from fifth street even though he has to walk to eighth street to the corn stand to actively trade his tomatoes for corn. The farther the store, the more efforts must be devoted. Costs of 1 effort per grid block, 2 efforts per grid block, and 3 efforts per grid block are tested.

3.4 Results

In this scenario, agents with the same composite goods as the second experiment are distributed over space, so that the cost of making a trade goes up in proportion to the traders' Manhattan distance in a 4×4 grid. Agents have two locations: a store, from which they hang their sign, if they choose to engage in passive trading, and the location of their buying agent, that travels around from store to store to make active trades. While death, which stressed the agents in [13], is logically more of a stress on the individual recognition treatments, space is more directly a stress on the role recognition agents. If individuals are ever to do better than roles, it seems that the scenario where they would do better would be a spatial scenario, where agents are too far apart to benefit from replacing each other. If there is too much space between a seller of barley on one end of the grid and a seller of barley on the other, how can one agent's transactions benefit from another's transactions?

In this experiment, three levels of spatial separation are studied, at a cost to get to an adjacent grid of 1 unit of effort, 2 efforts, and 3 efforts. These levels will be known as spatial-1, spatial-2, and spatial-3. The results show that these are hard on the agents, because the utility values of both the individual recognition treatments and the role recognition treatments decrease at a confidence greater than the 99% level between spatial-1 and spatial-2. The difference is not significant for between spatial-2 and spatial-3.

Figure 3 shows the results for the spatial scenario in tabular form. In spatial-1, the role recognition agent's utility is higher than the individual recognition agent's utility at the 99% confidence level, with individual recognition runs having an average utility of 139 while the role recognition runs have an average utility of 159. Average mutual information is better in the role treatment at above the 99% confidence level, with individual treatments averaging 0.26 for mutual information, and role recognition treatments averaging 0.71. Utility correlates with mutual information for the individual recognition treatment at 0.67, and for the role treatment at 0.48. The correlation is significant above the 99% confidence level.

In spatial-2, the role recognition agents still trade better the individual recognition agents, but at the 90% confidence level. The individual treatment has an

Treatment	Cost of Trade	Average Utility	Treatment Mutual Information	Correlation Utility and Mutual Info
Role	Spatial-1	159	0.71	0.48
	Spatial-2	139	0.28	0.79
	Spatial-3	140	0.12	N/A
Individual	Spatial-1	139	0.26	0.67
	Spatial-2	133	0.1	0.68
	Spatial-3	132	0.05	N/A

Fig. 3 Results for Spatial Scenario. Utility is higher in the role treatment than in the individual treatment. Role mutual information is higher than Individual mutual information as well

average utility of 133 while the role recognition treatment has an average utility of 139. The difference in average mutual information is significant at the 93% level, with the individual treatment having an average mutual information of 0.1, and the role treatment having an average mutual information of 0.28. The role recognition treatment has a drastically reduced amount of information that its symbol system can hold, as compared to spatial-1. However, the correlations between mutual information and utility are very strong, with a 0.68 for individual runs and a 0.79 for the role recognition runs. Both correlations occur at above the 99% confidence level. The difference in mutual information between the individual treatment and the role treatment is above the 99% level as well. This indicates that the mutual information is responsible for what little utility that the runs are able to muster.

Spatial-3 has no significant results with mutual information. The average mutual information of the individual treatments is 0.05 while the role recognition treatment is 0.12. This is the first experiment we have seen with no correlation between utility and mutual information. It means that the role recognition treatment did not use its symbol system to improve its utility over the individual recognition treatment. A look at the nature of the trades gives us a clue to why. Now that the agents are spatially arranged, we can look at pictures of what is going on.

We present a spatial illustration of the last day of trade in sample runs for both treatments in every cost of trade scenario. Most of the illustrations are of the median run for the role and the individual average mutual information. However, the best mutual information for the role and the individual treatments are included in Figs. 4 and 5, to illustrate what can happen. The best role treatment has an average utility of 164. The agents have a utility of 174 and a mutual information of 1.831 on the last day of this run. In these illustrations, the color of a square indicates an agent's sign, and the different shapes on the arcs between agents indicate the different goods that are traded. The shape located on an agent represents the good it receives in a trade. In this scenario, we have five different signs with exactly five different behaviors. Agents 2 and 4 both display the same sign (dark gray) and both have similar behavior. They buy good 2 and sell good 1 at a ratio of 3 to 2. Agents 1 and 9 both display the light gray sign, and have the same behavior. They buy composite good 7 and sell composite good 6 at a ratio of 4 to 3. Agent 3 buys the same goods and sells the same composite goods, but sells them at a different price, a ratio of 1 to 2. He has the medium gray sign. Agent 6 buys 2 and sells 3 at an even ratio, with a gray sign, and agent 0 buys 1 and sells 0 at a 3 to 2 ratio. Agent 3 is able to get another price for its goods because it is located distantly from the other stores that sold the same goods. It is the equivalent to a "convenience store". Note also how agents selling complex goods have to set up trade relations with agents selling their components (if they don't harvest the goods themselves). There is much to learn from utility, price and distance with SISTER.

Figure 5 shows the best individual treatment mutual information. It doesn't involve the complex goods as the best role treatment does. However, it does have two different signs with two different behaviors. The agent 2 (dark gray) sells 0 and buys 1, while agent 10 (light gray) buys 3 and sells 2.

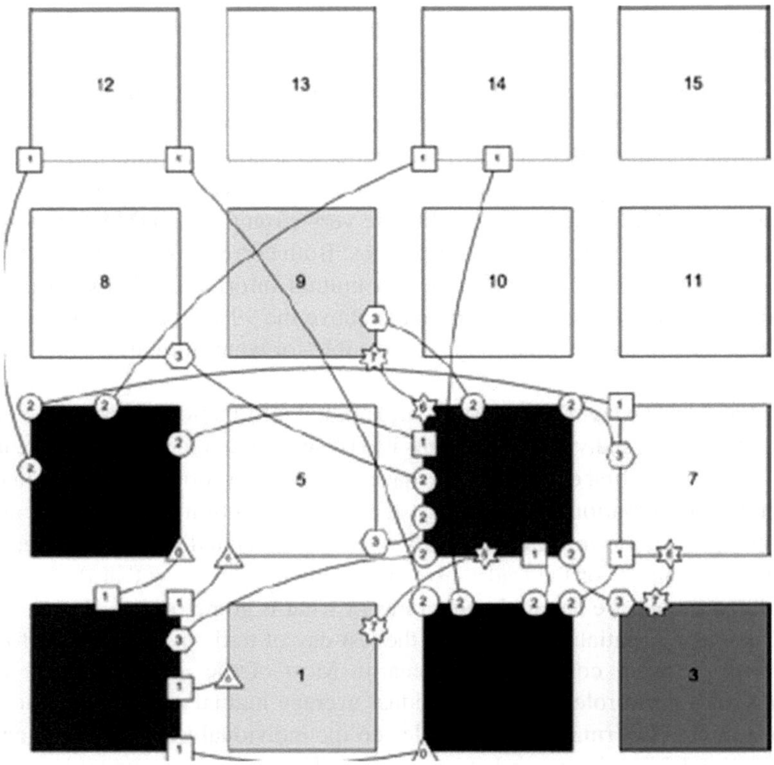

Fig. 4 Best mutual information in role recognition treatment for spatial-1. Agent "stores" are placed in fixed locations on a 4×4 grid. *Colors* represent different displayed signs. These agents have different behaviors for the displayed signs, meaning they have differentiated into roles. The *shapes* represent goods traded. In this run, both the *light gray* and the *medium gray* roles trade in the two composite goods, but the *medium gray* charges a different price and has different local customers than the light gray. Agents with complex goods have complex networks of trade needed to obtain the right components

Figure 6 shows the median run for the role treatment in the spatial-1 scenario. Its mutual information is 0.628, even though it actually has better utility than the best role treatment mutual information. This run's last scene utility is 186, while the best mutual information run's is 174.

The lower mutual information implies that some of the agent having different signs have some of the same behaviors, and some of the agents with the same sign have different behaviors. However, on the whole, signs have meanings. Most agents displaying the light gray and dark gray signs have the behavior of buying 3 and selling 2 at a 3 to 4 ratio. Most agents with light gray also buy 3 and sold 4 at a 3 to 4 ratio. Agents with the dark gray sign all buy 2 and sell 4 at a 2 to 1 ratio, while one of the agents with the dark gray sign and the gray sign agent buy 1 and sell 3 at a 1 to 3 ratio.

This is a very interesting scenario because it shows that one of the goods, number 3, has become the standard of trade. This explains why the mutual information is lowered

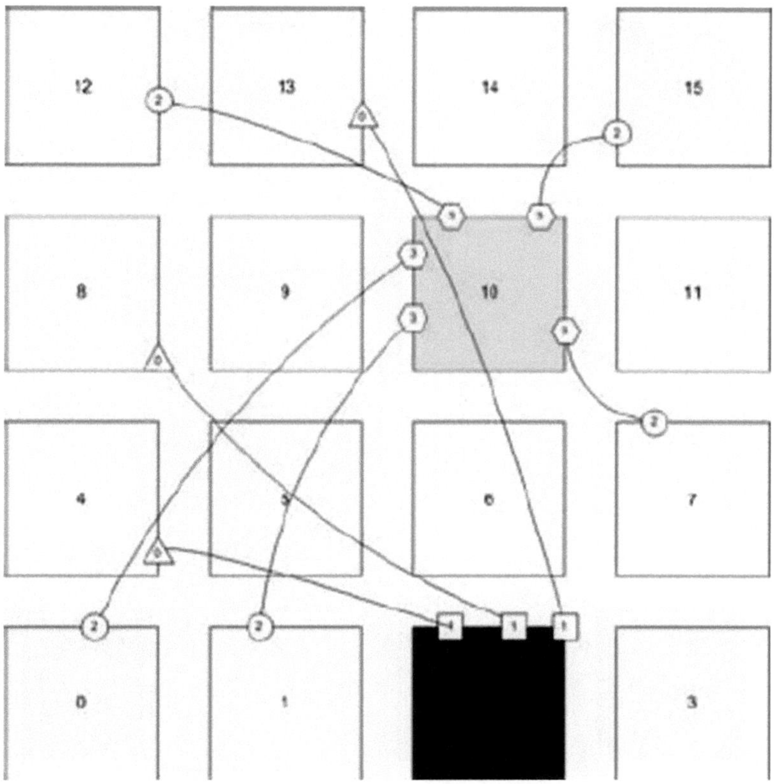

Fig. 5 Best mutual information in individual recognition treatment in spatial-1. Two of the agents have learned different roles, but they have not spread to other agents. No complex goods are traded

in such a high utility scenario. Money has emerged in this scenario. Agents use good 3 to get other goods that they want. But money is a strategy that lowers mutual information, because it means that agents, regardless of the sign they display, have the same behavior. It is not unique to this scenario, in fact, significant amounts of trades that get re-traded occur in 55% of the role composite good scenarios and in 35% of the individual treatments for the simple composite goods of the second experiment. These trades in goods that get re-traded, or exchange trades, indicate the emergence of money. It shows that SISTER, in keeping the principles of social science closely, has emerged many different institutions and has a lot to offer as an economics simulation.

Figures 7 through 11 are all scenarios of zero mutual information. However, the role society still does better than the individual society. This is because the role runs can still spread a single kind of store throughout the society. However, a single sign that means something has no mutual information, because it is not a system of signs that can mean different things. The commonness of zero mutual information in the higher cost of trade scenarios shows that these were severe stresses to both the individual treatments and the role treatments.

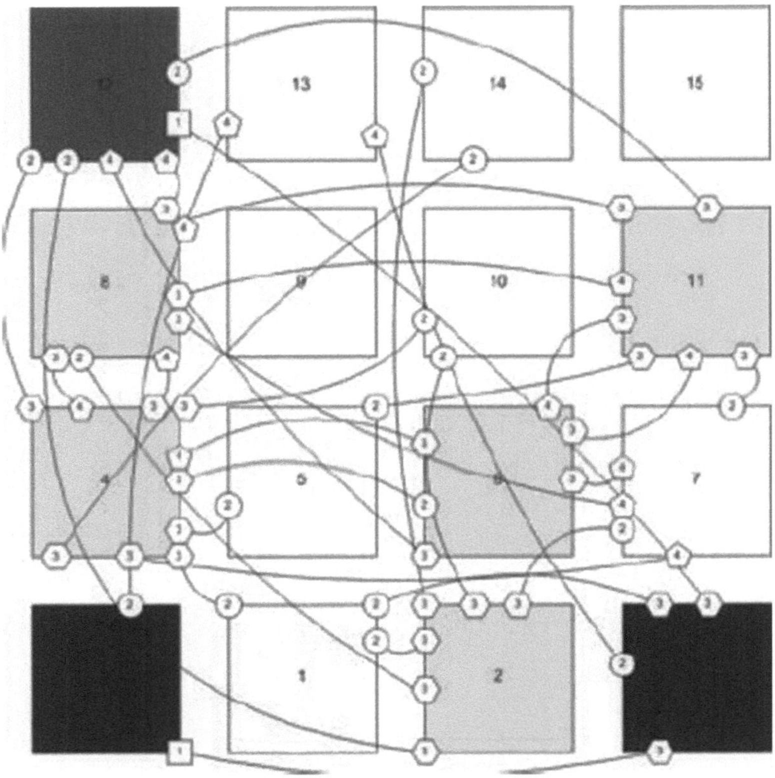

Fig. 6 Median mutual information in role recognition treatment in spatial-1. Three roles have formed. An interesting thing about this run is that good three has become a standard of trade, that is, money has emerged. Money emerges in 55% of the role scenarios. It makes the mutual information lower because agents trade in money regardless of their sign. Although this is the median mutual information run, it is the best utility run of spatial-1, perhaps because of the emergence of money

3.5 Discussion

This experiment shows that knowledge can be preserved over space in role recognition societies better than in individual recognition societies. A role allows adjacent agents to learn from each others experiences, and this causes information to spread over distances. This experiment, like that last one, demonstrates that knowledge in a role society can be preserved in the face of stress. Roles help cultural knowledge to have continuity over geographical distances as well as over generations.

SISTER is a study of the "free tags" of the original symbolic interactionist simulation of the emergence of social classes [6]. The free tags were the equivalent of words in a language, but applied to the identification of people. The dynamics involved in the emergence of meaning of tags are the same for the more general emergence of meaning of words. Symbolic interactionist simulation kept the principles

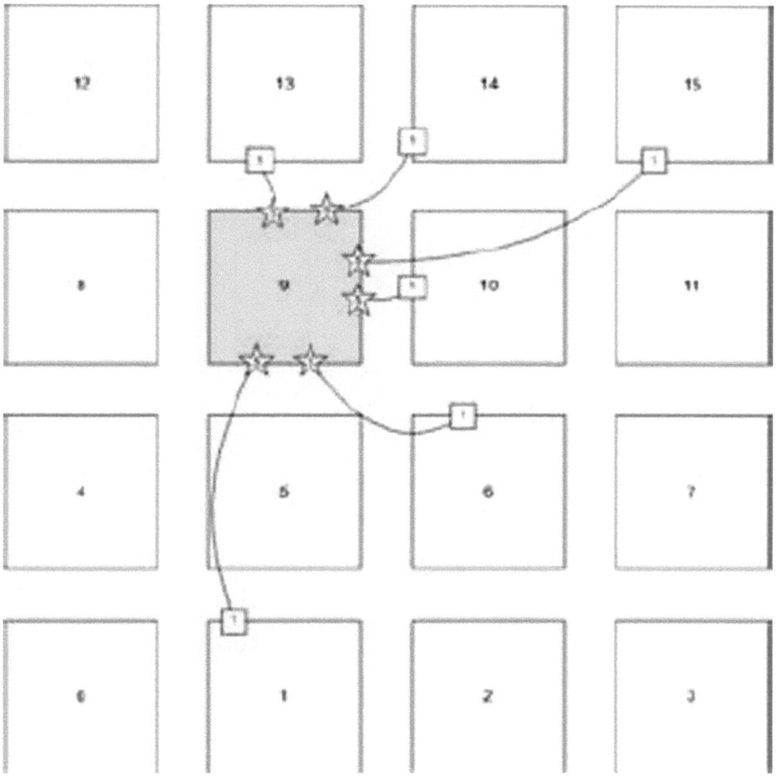

Fig. 7 Median mutual information (of zero) in individual recognition treatment of spatial-1

of autonomy and hermeneutics in its study of the emergence of language that subsequent more well known works, such as Steels', did. However, it also addressed critical issues that they did not. Steels and subsequent studies of the emergence of language are separated from studies of the emergence of culture. What is missing are models of language as coevolving with culture, models which capture the coevolutionary dialectic in which language and culture create each other and enable each other to grow. The dynamics of the propagation of signs which start out random is studied, but the dynamics of how they come to denote, hold, and spread new concepts needs more exploration. SISTER models the emergence of language as a dynamic creator of culture. If we define culture as the knowledge available to a society, both of the objects and the social structure, then SISTER shows how symbols emerge to hold culture and allow it to complexify, and how they enable culture to continue despite the deaths of individuals.

SISTER offers a solution to the hermeneutic paradox as do Steel's models, of how it is that people can only interpret the meaning of signs from the context of their individual life experiences, and yet still come to share meaning. SISTER agents are autonomous because they are closed with respect to meaning: they each have their own private induction mechanisms, and do not copy one another's signs

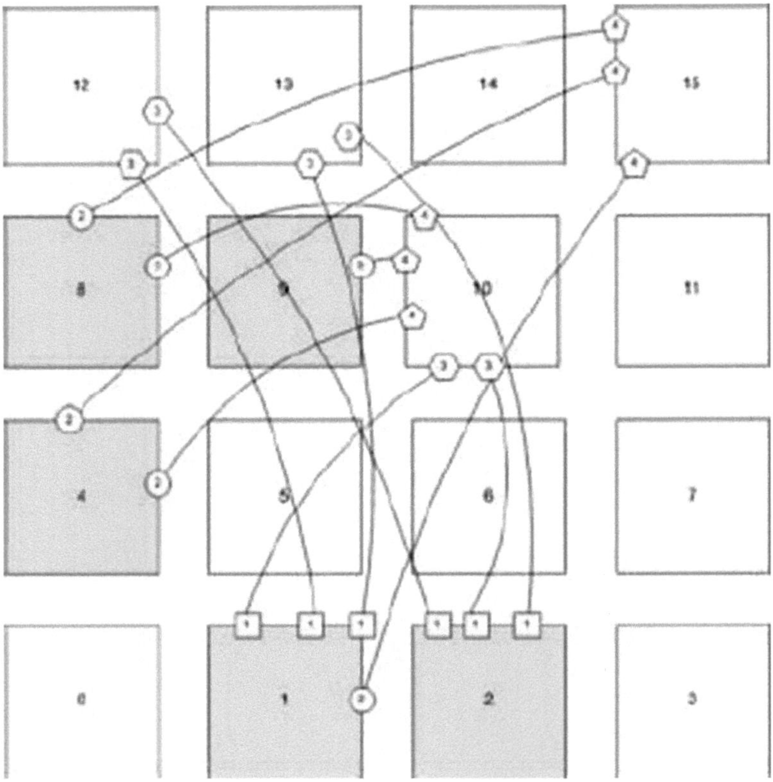

Fig. 8 Median mutual information (of 0) of role recognition treatment in spatial-2. Even though the mutual information is zero, it does better than the individual recognition treatment because many traders have taken on that one role

or interpretations of signs, but induce the meanings of the signs from their own experiences alone. SISTER however, is different from Steels' work in that the feedback is directly connected to the utility of the agent. A sign gets a particular interpretation based on what is good for the agent for it to mean, for its survival, rather than from the grunting approval of another agent. SISTER agents see "as the frog sees green" … just as the frog does not observe reality as it is, but constructs it as is beneficial to its survival [22], so do SISTER agents interpret signs based on whatever it is that gets them the most food. The combination of a direct relation of interpretation to utility along with perceptual autonomy is what makes SISTER agents both embodied and situated. If we do not model the advantage to utility that an interpretation confers at every step, we lose the ability to model important social processes of what becomes popular.

One example of such a process to model is that of the legend. Legends hold deep cultural meaning, often so deep as to be universal. Legends are told and retold orally over many generations. Each time they are retold, the teller contributes to the

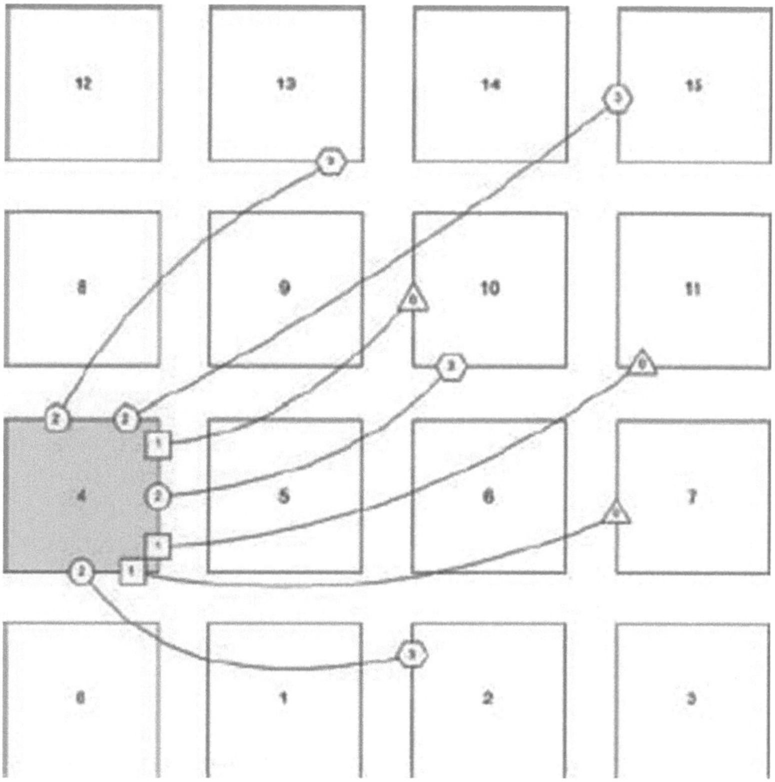

Fig. 9 Median mutual information (of 0) in individual recognition treatment of spatial-2

creation of the legend in small ways. As all the authors of a legend recreate it to meet their needs, it comes to be very good at meeting needs, settling down on a compromise between all needs. Imitation without such modification does not promote cultural products which contribute to the needs of all, deeply intertwined with the rest of the culture. It is not a deep consensus.

The principles of hermeneutics are important to the study of the emergence of language because we cannot separate language learning from concept learning, concept creation, and language creation. If we look at language as a passive thing, it does not matter if we include utility or not. If all a word is, is a random sign, and all we are explaining is how one random sign gets chosen over another random sign, then we need look no further than imitation. However, if we look at a word as a holder of a concept, a concept which serves to meet the needs of people within a web of other concepts, and which can only emerge as a word to denote it emerges, then it is appropriate to model the emergence of words in agents which interpret their meanings solely from their individual perspectives and usefulness to their lives. All the interpretations together create words and concepts which best serve the cultural needs of all the individuals. In the study of the emergence of language,

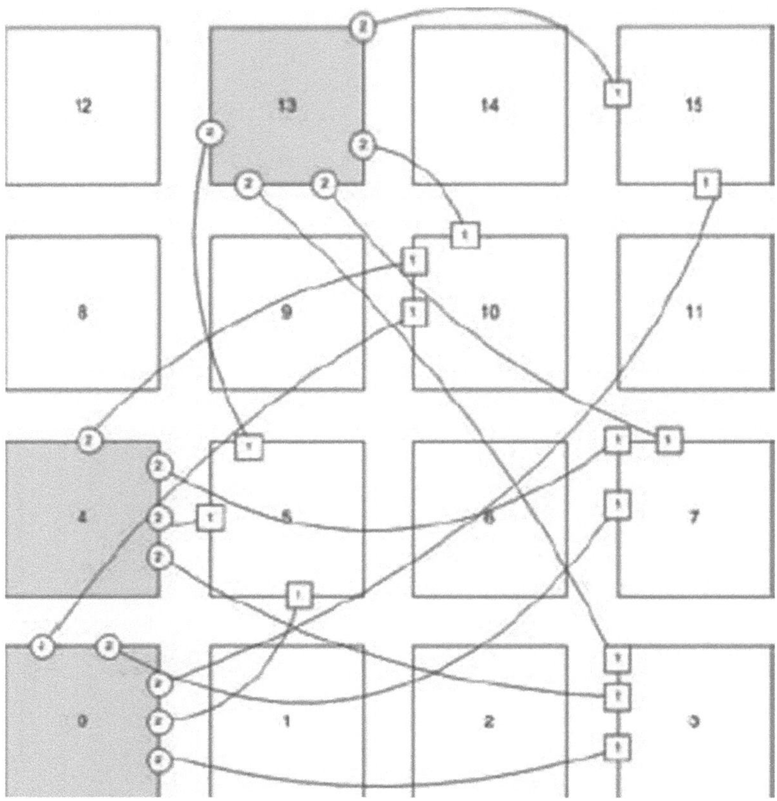

Fig. 10 Median information (of 0) of role recognition treatment spatial-3

it is not the sequence of phonemes that becomes popular that is important, but rather the capturing of the dynamic in which words make possible the ontologies that we use to construct our world. Studies in the emergence of language should address how words make the most practical ontologies, through the contributions of all utterers of words, rather than address the most practical sounds uttered.

SISTER shows that social systems with an emergent symbol system denoting an ontology of roles can enable cultural knowledge to continue despite the deaths of its individual members. The reason that it can continue is that signs denoting roles create expectations of behavior in agents who interact with a role. These expectations serve to train newcomers to the society into the proper behaviors of the role. Each sign for a role is a focal point of a set of social behaviors in a social network, in that the sign means a different thing to different other roles in a social network, and agents of each role have a certain set of expectation for agents of other roles that they interact with. The signs and the set of relations they denote are emergent, and must be emergent if they are going to denote any arbitrary set of behaviors. The knowledge in the society is held in the expectations that signs bring to the different agent's

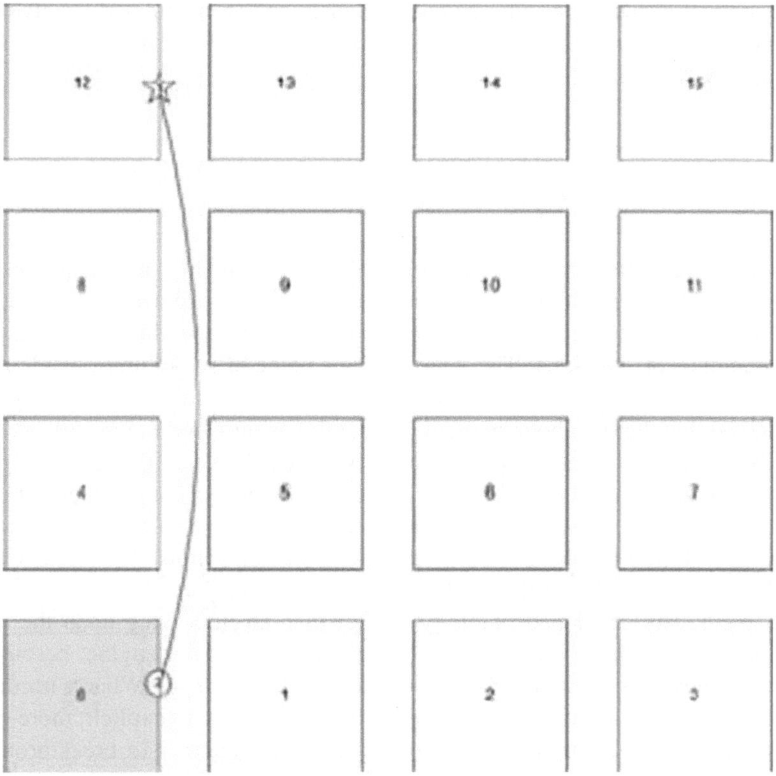

Fig. 11 Median mutual information (of 0) of individual recognition treatment of spatial-3

mind. These meanings are all induced by the private inductive mechanisms of agents, and yet the meanings of the signs come to be shared.

SISTER outputs a division of labor and social structure that increases the utility (that is, "satisfaction") of agents. Agent ontologies of roles emerge that guide agents in complex social relations and behaviors needed for survival. SISTER captures the fundamental social process by which macro-level roles emerge from micro-level symbolic interaction. SISTER comprises a multi-agent society in which agents evolve trade and communication strategies over time through the use of tags. The knowledge in SISTER is held culturally, suspended in the mutual expectations agents have of each other based on the signs (tags) that they read and display. Language emerges and is maintained robustly, despite the stresses of deaths of individual agents. SISTER shows how a complex endogenous communication system can develop to coordinate a complex division of tasking.

SISTER employs coevolution, in which agents each have their own genetic algorithm (GA), whose fitness is dependent on successful interaction with other agents. These GAs evolve tags that come to indicate a set of behaviors associated with a role. Roles are nowhere determined in the simulation and exist in no one place, but rather are suspended in the mutual expectations of the coevolving agents.

These mutual expectations emerge endogenously and are expressed through signs with emergent meanings. All institutional knowledge is distributed in these subtle mutual expectations.

4 Future Directions

When Desalles et al. praised Axtell et al.'s strong (symbolic interactionist-style) emergence, Desalles et al. noted an immergence, a downward irreducible causation that changed the behavior of the races by means of a tacit, rather than an explicit, understanding of the signs. The signs did not point to something outside of the agent, they point to utility alone as in Maturana et al.'s frog that sees green. Desalles et al. noted that the (symbolic interactionist-style) agent's internal models were not reflexive, that they did not map to the agent's world. However, Desalles along with many other current theorists of "immergence" fail to realize that it is the tacit nature of the model that allows an entire social engine to form, an invisible hand that makes need-filling institutions out of individual selfish actions. Desalle et al. proposed an improvement to Axtell et al. in which agents can categorize their knowledge into a previously developed ontology. Rather than an improving upon the strong emergence this change would disable the autonomous social engine, because the previously developed ontology is an exogenous and static input. What is needed for true objectivity, the move from tacit as-the-frog-sees-green to explicit, more objective models of the environment that is entirely endogenous is a breakthrough in cognitive science. Since endogenous objectivation is beyond our technical knowledge, tacit knowledge is the only simulatable phenomena that can form an entire need filling engine at this time.

Of course, people cognate detailed models of the environment for their utility just as Maturana et al.'s frog did, and even though no one person has a complete explicit map of the entire world of thought, these models are more shared than the tacit knowledge of Maturana et al.'s frog. This objective knowledge is useful in society and to the symbolic interactionist practice of "taking the shoes of another." In order to go from one's own selfish functional perspective to someone else's selfish functional perspective, one must pass through an objective "hub" to see other functional "spoke" points of view first. The technology that could put an agent in the shoes of another would be a technology that could take in correlations that it an agent discovered through functional induction, and put out an objective model of cause. Until cognitive science is at the point where it can derive an objective causal simulation from subjective correlative data, programs which purport to simulate immergence must use tacit models. The alternative, considering the state of the science now, is to hard code a representation of the "emergent" property, losing the endogeny necessary for the simulation's fidelity. In the mean time, it is best to, as Holland did, recognize that a tacit model is just as much an internal model as an explicit model.

Endogenously created cognitive maps would go a step farther in simulating the symbolic interactionist paradigm, as reflexivity at the level of getting into the other's shoes is required, and thus the ability to find an objective representation is needed. Further, symbolic interactionist simulations to this point have only covered the first two mechanisms in Holland's recipe for complex adaptive systems: tags and internal models. They have no building blocks, no dynamically recombinable signs that can mean new things to be interpreted during the interaction, as in Garfinkel's ethnomethodology in symbolic interactionism requires [15]. Endogenous internal causal models from correlated relations and recombinable symbols that are in language are ambitious next steps for not only the symbolic interactionist paradigm, but for cognitive science in general. Maybe the techniques of cognitive science can benefit from the techniques of symbolic interactionism in these next steps for modeling emergent meanings.

References

1. Axtell R, Epstein J, Young H (2001) The emergence of class norms in a multi-agent model of bargaining. In: Durlouf, Young (eds) Social dynamics. MIT Press, Cambridge
2. Bedau M (2002) Downward causation and the autonomy of weak emergence. Principia 6(1):5–50
3. Berger P, Luckmann T (1966) The social construction of reality. Anchor Books, New York
4. Coleman J (1994) Foundations of social theory. Belknap, New York
5. Dessalles JL, Müller JP, Phan D (2007) Emergence in multi-agent systems: conceptual and methodological issues. In: Phan D, Amblard F (eds) Agent based modelling and simulations in the human and social sciences. The Bardwell Press, Oxford, pp 327–356
6. Duong DV (1991) A system of IAC neural networks as the basis for self organization in a sociological dynamical system simulation. Masters thesis, The University of Alabama at Birmingham. http://www.scs.gmu.edu/~dduong/behavior.html
7. Duong DV, KD Reilly (1995) A system of IAC neural networks as the basis for self organization in a sociological dynamical system simulation. Behav Sci 40(4):275–303. http://www.scs.gmu.edu/~dduong/behavior.html
8. Duong DV (1995) Computational model of social learning. Virtual School ed. Brad Cox. http://www.virtualschool.edu/mon/Bionomics/TraderNetworkPaper.html
9. Duong DV (1996) Symbolic interactionist modeling: the coevolution of symbols and institutions. Intelligent systems: a semiotic perspective. In: Proceedings of the 1996 international multidisciplinary conference, vol 2, pp 349–354. http://www.scs.gmu.edu/~dduong/semiotic.html
10. Duong DV (2004) SISTER: a symbolic interactionist simulation of trade and emergent roles. Doctoral dissertation, George Mason University, Spring
11. Duong DV, Grefenstette J (2005) SISTER: a symbolic interactionist simulation of trade and emergent roles. J Artif Soc Soc Simulat. http://jasss.soc.surrey.ac.uk/8/1/1.html
12. Duong DV, Grefenstette J (2005) The emulation of social institutions as a method of coevolution. In: GECCO conference proceedings. http://www.scs.gmu.edu/~dduong/gecco.pdf
13. Duong DV (2009) The generative power of signs: tags for cultural reproduction. In: Trajkovsky G, Collins SG (eds) Handbook of research on agent-based societies: social and cultural interactions. IGI Global, New York
14. The Economist (1997) What boys and girls are made of, p 96. http://www.scs.gmu.edu/~dduong/economist.pdf
15. Garfinkel H (1967) Studies in ethnomethodology. University of California, Los Angeles
16. Hales D (2004) Tags for all! – understanding and engineering tag systems. In: 4th International conference on complex systems (ICCS 2004). Springer, New York
17. Hodgson G (2006) What are institutions. J Econ Issues 15:1

18. Holland JH (1975) Adaptation in natural and artificial systems. University of Michigan Press, Ann Arbor
19. Holland J (1993) The effects of labels (tags) on social interactions. In: Sante Fe Institute Working Papers. The Santa Fe Institute, Santa Fe
20. Lacobie KJ (1994) Documentation for the Agora. Unpublished document
21. Luhmann N (1984) Social systems. Suhrkamp, Frankfort
22. Maturana H, Lettvin J, McCulloch W, Pitts W (1960) Anatomy and physiology of vision in the frog. J Gen Physiol 43:129–175
23. Muller J (2004) The emergence of collective behavior and problem solving. In: Agents world IV international workshop. Springer, New York, pp 1–20
24. Parsons T (1951) The social system. Free Press, New York
25. Shannon CE (1993) Collected papers. Wiley. New York
26. Steels L (1996) Emergent adaptive lexicons. In: Fourth international conference on simulation of adaptive behavior, Cape Cod. Springer, New York
27. Winograd T, Flores F (1987) Understanding computers and cognition. Addison-Wesley, New York

Virtual City Model for Simulating Social Phenomena

Manabu Ichikawa, Yuhsuke Koyama, and Hiroshi Deguchi

Abstract In our research, we construct a virtual city model which is a base tool for simulating a social phenomenon that will occur in a city. This model supports a plug-in system. In this system, a social phenomenon is required to be constructed as a plug-in module. A social phenomenon simulation model can be made if a social phenomenon module plugs in this virtual city model. In this model, it requires data which include the number of homes, shops, schools and etc in the real world and enables to represent a real city as a virtual city. By changing data, a variety of cities can represent as virtual cities. An example of a simple virtual city model using this model and also a social phenomenon model, an infection spreading model, are shown in this paper.

Keywords City simulation • Virtual city • Agent based simulation • SOARS • Social simulation

1 Introduction

In recent years, researches which use multi agent systems become active, especially in the range of the economic phenomenon, the traffic flow simulation, the infection spreading simulation and so on. Researches which target the human society and social phenomena increase [1–3]. However, to simulate these things, they usually require a virtual environment [5]. For example, to simulate the traffic flow, it needs a virtual city which contains information of roads in the city. To make

M.Ic hikawa(✉), Y. Koyama, and H. Deguchi
Department of Computational Intelligence and Systems Science, Tokyo Institute of Technology, 4259N agatsuta-cho,M idori-kuY okohama,226- 8502,J apan
e-mail: ichikawa@dis.titech.ac.jp; koyama@dis.titech.ac.jp; deguchi@dis.titech.ac.jp

K. Takadama et al. (eds.), *Simulating Interacting Agents and Social Phenomena: The Second World Congress*, Agent-Based Social Systems 7,
DOI 10.1007/978-4-431-99781-8_18, © Springer 2010

such a virtual city, usually both Geographic Information Systems (GIS) and a Cellular Automaton (CA) are used. By using this methodology, as GIS uses map data of the real world, models that target movements of vehicles and human walking or models that target the change of the population and the migration can be easily made and it is suitable for simulating the change of social phenomena on the map. But it is not suitable for simulating some social phenomena that contain human interactions which are done in a building, for example, a simulation of how a rumor is spread in a city. Because in these simulations, map data of the real world is not important and to represent places that humans can interact with others are important. So, a new type of a virtual city that does not use GIS and CA is needed. In this research, we construct a virtual city model which is for simulating a social phenomenon that will occur in daily life in the real world and that does not need map data of the real world. In this model, places where human can interact with others or do some activities are represented according to data of the real world and any types of cities can be represented as if there are data for setting this virtual city. It is designed to construct a virtual city model automatically by just reading data. A social phenomenon which is going to simulate should be designed as a plug-in module for this virtual city model. By preparing a social phenomenon module on this virtual city model, it is possible to simulate a social phenomenon which will occur or occurs in a city and it becomes a model for simulating a social phenomenon. As this model is made by using simulation language SOARS, users can design modules and reconstruct a model not only by SOARS but also by Java [4]. In this paper, we introduce structures of this virtual city model and show an example of plug-in module.

2 Abstract of SOARS

SOARS is designed to describe agent activities under the roles on social and organizational structure. Role taking process can be described in our language. SOARS is also designed depending on the theory of agent based dynamical systems. Decomposition of multi-agent interaction is most important characteristics in our framework. The notion of spot ant stage gives special and temporal decomposition of interaction among agents [7].

New information of SOARS can be gotten from <http://www.soars.jp/>.

3 Details of the Model

In this section, we explain details of the model and introduce an example of a virtual city model which uses an architecture that will explain in this section [6].

4 Structure of Virtual City

Most of models that were aimed to model a town or a city use the geographic information, usually GIS (Geographic Information System), for modeling. Using this methodology is useful for modeling a very narrow area in the real world but it is difficult to model a very huge area, for example, whole city, because data for modeling become very huge and complex. In this virtual city model, instead of using the geographic information, the layer information is used. In the layer information, the structure of the city is represented with two or more layers. In this virtual city model, the structure of the virtual city is represented with four layers, City Layer, Zone Layer, Area Layer and Building Layer, which are shown in Fig. 1. In this virtual city, as there is no using of information of geographical data, the movement of an agent is expressed by moving between layers. For example, if an agent who is in home which is located in Area1, Zone1 wants to go to a building which is located in Area6, Zone2, it moves like follows: Home=>Area1=>Zone1=>City=>Zone2=>Area6=>Building.

5 Settings for the Virtual City

To define a virtual city, the number of zones, areas and buildings are required. All these data are filled in a data sheet, an example of this data sheet is shown in Table 3, and a data sheet is read when a virtual city starts simulation and a layer

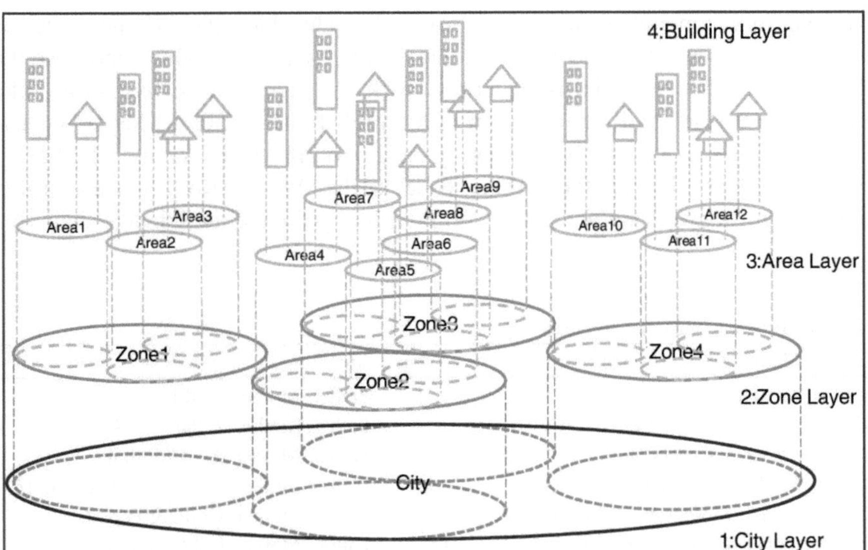

Fig. 1 Layer structure of virtual city model

structure of a virtual city is constructed automatically. This data sheet enables for users to make a virtual city model which is similar to a real city by using a real data of the real world or which is based on user's assumption.

In this virtual city, there are 17 types of homes and the number of each type of home is defined in a data sheet. Differences of areas are represented by this definition. For example, an area where a lot of people live alone, an area where many children live or an area where a few people live can design in the same virtual city. Types are shown in Table 1.

6 Behaviors of Agents in the Virtual City

In this model, behaviors of agents depend on roles that agents have. A role is a bundle of rules in SOARS and all rules of behaviors are defined in one of all roles. Agents always see a role which is active and act according to rules which are defined in it. Agents are also able to change their roles according to their situation, so they can act complex actions by combining roles.

There are three types of roles in this model. These roles are simple but they become base of other roles which are defined by users. The first one is based on domestic roles and the second one is based on social roles. The third one is based on general roles. Domestic roles mean roles in a home, Social roles mean roles in a social life and general roles mean roles which agents generally have in common. We show types and roles, and their meanings in Table 2.

We explain details of behaviors of agents who have the "Mother Role" as an example. In this model, agents, whose sex is "Female", have the "Mother Role" as a domestic role whether they live with their family or they live alone and have the "Worker Role" as a social role if they work. All agents who live alone take the "Worker Role" and others take it at the probability of 50%. This is because about 50% of women who have a family have a job in Japan. Agents who have a job change their role from the "Mother Role" to the "Worker Role" in the morning and they go for working and after working, they go home and change their role

Table 1 Type and meaning of home

Type	Meaning	Type	Meaning
A	Single(1)	J	Father, Mother, Child & Parent(4)
B	Father & Mother (2)	K	Father, Mother & Parents (4)
C	Father & Child (2)	L	Father, Mother & 3 Children (5)
D	Mother & Chile (2)	M	Father, Mother, 2 Children & Parent (5)
E	Father, Mother & Child (3)	N	Father, Mother, Child & Parents(5)
F	Father & 2 Children (3)	O	Father, Mother & 4 Children (6)
G	Mother & 2 Children (3)	P	Father, Mother, 3 Children & Parent (6)
H	Father, Mother & Parent (3)	Q	Father, Mother, 2 Children & Parents (6)
I	Father, Mother & 2 Children (4)	(Parent means Grandfather or Grandmother)	

Table 2 Types and roles, and their meanings

Type	Role	Meaning
Domestic Role	Father Role	Rules for a father in a home
	Mother Role	Rules for a mother in a home
	Boy Role	Rules for a boy in a home
	Girl Role	Rules for a girl in a home
	Grandfather Role	Rules for a grandfather in a home
	Grandmother Role	Rules for a grandmother in a home
Social Role	Worker Role	Rules for a person who works
	Student Role	Rules for a boy or a girl who goes to school
	Customer Role	Rules for a person who goes shopping
	Free Role	Rules for an agent who takes a walk
General Role	Human Role	Rules for all agents
	Adult Role	Rules for an agent who has a father role or a mother role as a domestic role
	Older Role	Rules for an agent who has a grandfather role or a grandmother role as a domestic role
	Child Role	Rules for an agent who has a boy role or a girl role as a domestic role

from the "Worker Role" to the "Mother Role". Agents who do not have a job remain their home and change their role from the "Mother Role" to the "Customer Role" if necessary and go shopping. After shopping, they go home and change their role to the "Mother Role". According to what agents should do they appropriately change their role.

- Mother Role:
  - 06:00 Agents who have a job change their role to the "Worker Role"
  - 09:00–15:00 Agents change their role to the "Customer Role" and go shopping at a certain probability
- Worker Role:
  - 07:00–09:00 Agents start to go to their office
  - 09:00–21:00 Agents work till the end time of working comes
  - 18:00–21:00 Agents start to go home if they have finished their work
  - 18:00–21:00 Agents change their role to a domestic role, the "Father Role" or the "Mother Role", if they arrived at their home
- Customer Role:
  - Anytime Agents start to go to a shop for shopping
  - Anytime After shopping, agents go home and change their role to a domestic role if they have arrived at their home

In this model, all agents who have the "Father Role" as a domestic role have a job and take the "Worker Role" as a social role and go to an office for working in daytime and all agents who have the "Boy Role" or the "Girl Role" take the "Student Role" as a social role and go to school in daytime. Agents who have the "Grandfather Role" or the "Grandmother Role" do not have a job and take the "Customer Role" or the "Free Role" if necessary.

It is true that those roles are not enough for representing the daily life in the real world and there are many other activities and there should be many other roles in this model. To make this model to have more reality, it is required to make more domestic roles and social roles. We suppose users to make more roles which are based on those basic roles to simulate a social phenomenon that users are interested in.

7 An Example of the Virtual City

We show an example of a virtual city of this model in this sub section. This virtual city is very simple and is not based on a real data. It is simplified for an example. Settings for this simple virtual city are shown in Table 3.

The total number of houses in this virtual city is 200 and 440 agents live in. Forty agents whose sex is "Male" and 40 agents whose sex is "Female" live alone and have a job. All Other 360 agents have their family. Hundred and twenty of 360 agents who have the "Child Role" for a general role go to the elementary school or the junior high school. A half of 240 agents of the remainder have the "Father Role" and others have the "Mother Role" as a domestic role and all agents who have the "Father Role" have a job and about a half of agents who have the "Mother Role" also have a job. To simplify this virtual city, there are no older agents which have the "Older Role" as a general role. Figure 2 shows the initial state of this virtual city model by animation.

Figure 3 shows the daytime state. Agents who have a job go to their office, agents who have the "Child Role" as a general role go to school and some agents go shopping.

8 Use Case of the Model

In this section, we show an example of using this virtual city model. As it is described in Sect. 1, all social phenomena which are required to represent in this virtual city can be achieved as plug-in modules of this model. Modules such as follows can be considered to be represented in this model.

- The model which represents customer's behaviors like Huff Model
- The model which represents how an infection spreads in a city
- The model which represents how an effect or a policy influences in a city
- The model which represents how rumor spreads in a city

As it is described before, it is impossible to simulate any kinds of social phenomenon by using this simulation environment. Because of not using CA architecture, it is

Table 3 The data sheet of simple virtual city

Zone	Station	Area	Type of building	Num
Zone1	Station1	Area1	Home TypeA	20
			Home TypeB	10
			Home TypeE	10
			Home TypeI	10
			Elementary School	1
			Junior High School	1
			Shop	2
			Office	5
		Area2	Home TypeA	20
			Home TypeB	10
			Home TypeE	10
			Home TypeI	10
			Elementary School	1
			Shop	2
			Office	5
Zone2	Station2	Area3	Home TypeA	20
			Home TypeB	10
			Home TypeE	10
			Home TypeI	10
			Elementary School	1
			Junior High School	1
			Shop	3
			Office	10
		Area4	Home TypeA	20
			Home TypeB	10
			Home TypeE	10
			Home TypeI	10
			Elementary School	1
			Shop	3
			Office	10

hard for this virtual city environment to simulate social phenomena which contain movement of humans or vehicles with selection of routes. In this virtual city environment, social phenomena that contain human interactions in buildings and area and that movement of humans or vehicles are not so important, for example, a social phenomenon that contains spreading of information and so on, are easily simulated. We assume for users to use this virtual city environment for simulating these kinds of social phenomena.

We show an infection model as an example use of this virtual city model. Algorithms of spreading an infection are defined in an infection module and agents act both infection behaviors and normal behaviors in the virtual city. Algorithms of spreading an infection are as follows:

1. An agent who has an infection and whose infection is in progress pollutes the place where it is now

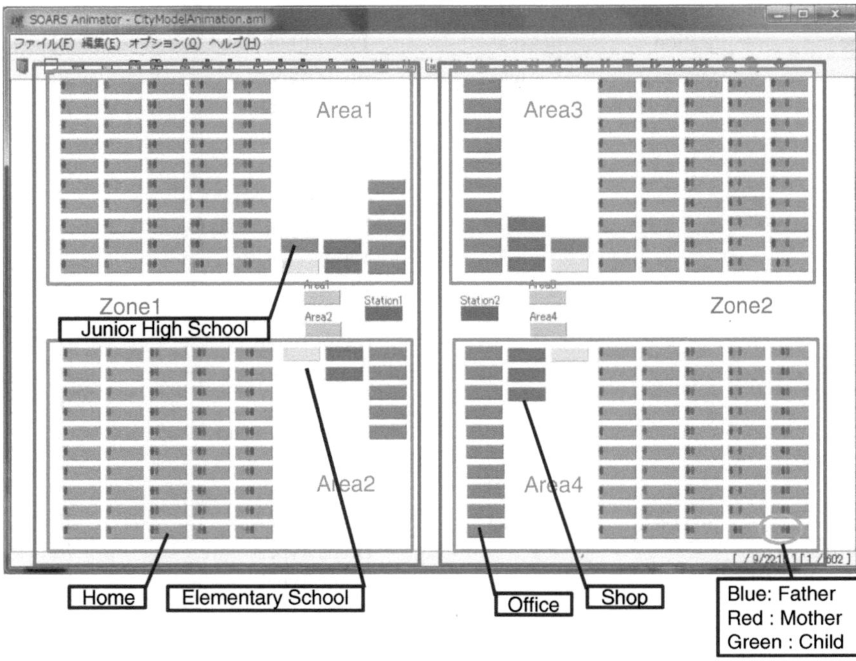

Fig. 2 Initial state of virtual city

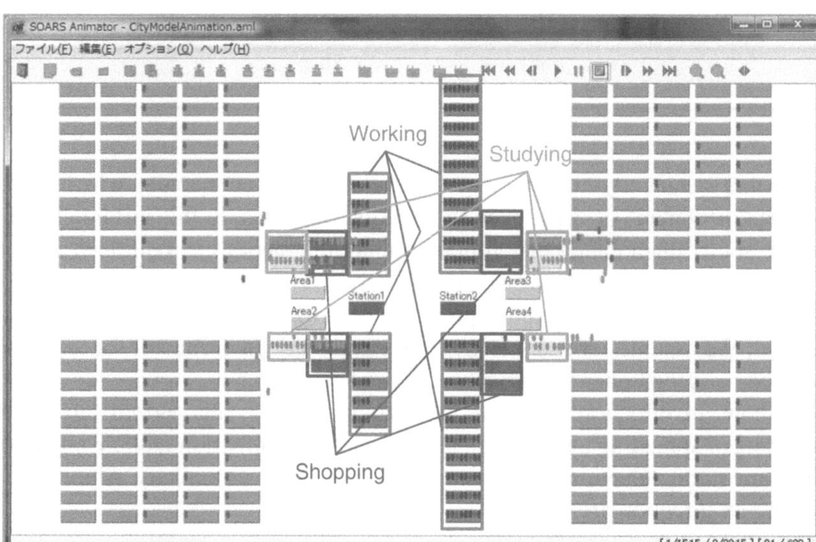

Fig. 3 Daytime state of virtual city

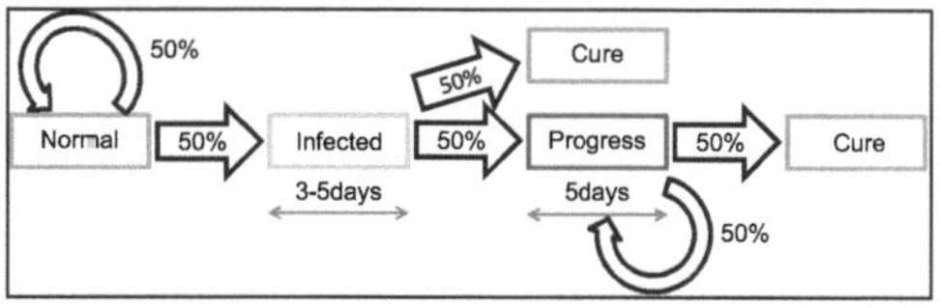

Fig. 4 Infection state

2. If an agent who does not have an infection is in a polluted place, it gets an infection at a certain probability
3. An agent who has an infection updates the state of an infection according to a flow chart of an infection state

A flow chart of an infection state is shown in Fig. 4.

From an infection state in Fig. 4, an agent who gets an infection change its condition from "Normal" to "Infected" at the probability of 50% and an agent who does not change its condition has a possibility to infect later. It means that once an agent gets an infection, all agents who have an experience of getting an infection will be infected. After 3–5 days from infecting, some agents cure their infection and others change their condition from "Infected" to "Progress". The probability of changing their condition is 50% in this infection state. Agents who cure their infection never get infection again. And after 5 days, agents whose condition is "Progress" cure their infection at the probability of 50%. Some agents who do not cure their infection in 5 days will cure in couple more days by using the 5th day cure algorithm. Figure 5 shows an example of this infection model. It shows the 9th day after one agent getting an infection and it begins to spread its infection in simple virtual city which includes this infection module.

From this kind of an infection model, it enables to see and know how an infection spreads in a city. And by remaking an infection module based on some other kinds of infections, it enables to know the difference of the way of spreading. If there are some virtual cities, it is also possible to know the difference of the way of spreading according to types of cities, for example, the difference between a metropolitan area and a bed town area can be known. And from the view of people, who are going to prevent this kind of an infection, results of this kind of model may be useful and can be used for making preventing policies. If these people make some modules of preventing infection policy, they will be able to know effects of a policy and they will also be able to compare policies by using both an infection module and a preventing policy module in the virtual city.

9 Future Works

In this paper, we show a simple virtual city and an example of using this virtual city model. In Japan, the government takes a national census every 5 years and it is possible to estimate the number of houses according to the structure of family, the

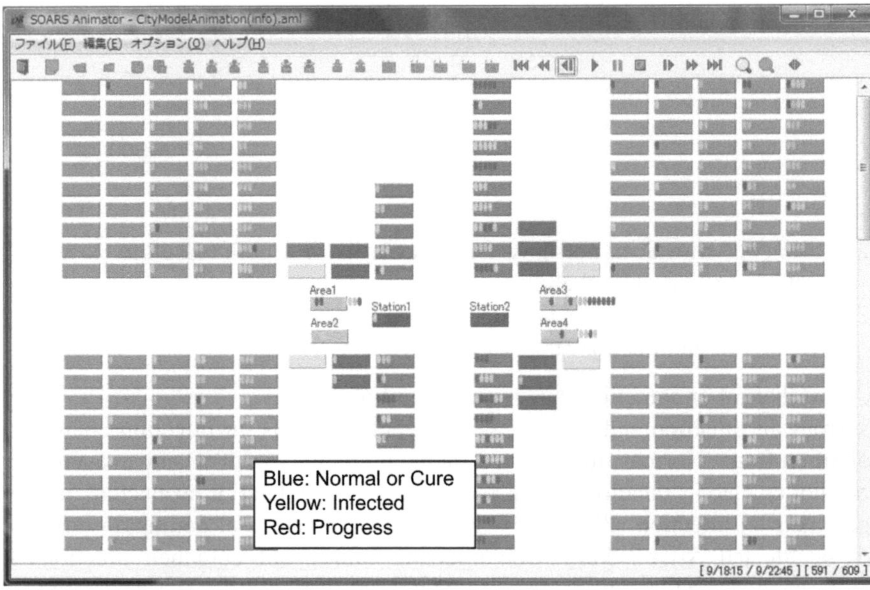

Fig. 5 Ninth day of virtual city from infection starting to spread

number of living people and the number of houses which are in the same area from its census. A data sheet of a virtual city model, which is shown in Sect. 2, can be filled by using these estimating data and it enables to make any kinds of Japanese virtual cities or towns which are based on a real data. If there are same kinds of data in other countries, it is also possible to make virtual cities or towns which are similar to the real world. At this present stage, as there is no software which makes the data sheet of a virtual city model from a national census automatically, users of this virtual city model have to estimate a national census for making a data sheet. However, it takes much time for estimating, so we are going to make software for making a data sheet from a national census automatically in the next stage. And also prepare some documents for making plug-in modules and for defining movement rules of agents for this virtual city model. If we prepare these documents, it will become possible for users to construct models in which agents act according to rules that are based on ideas of users and that are aimed to simulate a social phenomena that users are interested in.

References

1. Epstein J, Axtell R (1996) Growing artificial societies. MIT Press, Cambridge, MA
2. Gilbert N (2008) Agent-based models. SAGA Publications, London
3. Gilbert N, Troitzsch K (1996) Simulation for the social scientist. Open University Press, Philadelphia

4. Ichikawa M, Tanuma H, Koyama Y, Deguchi H (2007) SOARS for simulations of social interactions and gaming. Introduction as a social microscope. In: Proceedings of the 38th annual conference of the international simulation and gaming association, p 36
5. Ichikawa M, Koyama Y, Deguchi H (2005) A basic simulation model for evaluating social phenomenon. In: Proceedings of the fourth international workshop on agent-based approaches in economic and social complex systems (AESCS), pp 250–261
6. Richiardi MG, Leombruni R, Saam NJ, Sonnessa M (2006) A common protocol for agent-based social simulation. J Artif Societ Social Simulat 9(1):15. Available at SSRN: http://ssrn.com/abstract=931875
7. SOARS Project. http://www.soars.jp

Modeling Endogenous Coordination Using a Dynamic Language*

Jonathan Ozik and Michael North

Abstract Dynamic languages are computer languages that allow programs to substantially restructure themselves while they are running. Interest in these kinds of programming languages has dramatically increased in the last few years. This paper builds on previous work by exploring the use of a popular dynamic language, namely Groovy, within the Repast Simphony (Repast S) platform. This language is applied to modeling the endogenous emergence of coordination within a group of social agents. This paper introduces the Endogenous Emergence of Coordination (EndEC) model. It then highlights many of the features in Groovy that were found to be particularly helpful during model implementation. This demonstrates the powerful and flexible capabilities that a dynamic language can bring to the creation of agent-based models. What is particularly exciting is the potential for creating truly dynamic and evolving open-ended simulations, where the simulation fundamentally changes as it executes.

Keywords Dynamic language • Endogenous • Coordination • Emergence • Agent-based modeling

---

*The submitted manuscript has been created by UChicago Argonne, LLC, Operator of Argonne National Laboratory ("Argonne"). Argonne, a U.S. Department of Energy Office of Science laboratory, is operated under Contract No. DE-AC02-06CH11357. The U.S. Government retains for itself, and others acting on its behalf, a paid-up nonexclusive, irrevocable worldwide license in said article to reproduce, prepare derivative works, distribute copies to the public, and perform publicly and display publicly, by or on behalf of the Government.

J. Ozik and M. North
ArgonneN ationalLa boratory, Argonne,I L60439 ,U SA
Universityof C hicago, Chicago,I L60 637,U SA
e-mail: ozik@anl.gov, north@anl. gov

K. Takadama et al. (eds.), *Simulating Interacting Agents and Social Phenomena: The Second World Congress*, Agent-Based Social Systems 7,
DOI 10.1007/978-4-431-99781-8_19, © Springer 2010

1 Introduction

As previously discussed in Ozik et al. [9] and Ozik and North [8], dynamic languages have gained in popularity in recent years. Groovy [2,4] is a dynamic language that is tightly integrated with Java and, hence, it has a natural ability to integrate into the Repast Simphony [10] agent modeling platform. Repast Simphony (Repast S) is the latest extension to the Repast portfolio, a widely used, free, and open-source, agent-based modeling and simulation (ABMS) toolkit [6,7]. Repast S offers a variety of approaches to simulation development and execution, and includes many advanced features for agent storage, display, and behavioral activation, and new facilities for data analysis and presentation. This paper builds on Ozik et al. [9] and Ozik and North [8] by exploring in more detail the use of a dynamic language, Groovy, for providing the Repast S platform with support for modeling the endogenous emergence of coordination.[1]

After summarizing the Endogenous Emergence of Coordination (EndEC[2]) model, we highlight many of the features in Groovy that we found particularly helpful in building the model. Our intent is to demonstrate the powerful and flexible capabilities that a dynamic language can bring to the creation of agent-based models. What is particularly exciting is the potential for creating truly dynamic and evolving open-ended simulations, where the simulation fundamentally changes as it executes.

2 Related Work

The EndEC sample model is at least partially inspired by Holland's work on the emergence of language [3,13]. Tao et al. [13] state the following:

> In this model, language is treated as a set of mappings between meanings and utterances (M-U mappings). Communication is a language game, in which, agents, based on linguistic and nonlinguistic information, produce and comprehend utterances that encode with integrated meanings. The emergence of a common language is indicated by sharing a common set of rules among agents through iterative communications.

The model presented here uses the same approach of conceptualizing language as "a set of mappings between meanings and utterances" [13]. However, the model presented in this paper uses observable motion of agents rather than linguistic rules to define meanings.

[1]It is important to note that Repast S and its related tools are currently under development. This paper presents the most current information at the time it was written. However, changes may occur before the next release.

[2]A play on words, as an 'endec' in electrical systems terminology is a single integrated device which is both an encoder and a decoder for signals which, as will be shown, relates to the agent behaviors in the EndEC model.

In terms of the use of dynamic languages for agent modeling, Swarm [12] was originally designed to use Objective-C's dynamic method call binding to recognize and translate undefined method calls (i.e., 'selectors') [1,5]. In the next sections we will show how this is done in Groovy. However, unlike Groovy, Objective-C does not provide an ability to actually modify agent class definitions after compilation.[3]

3 The EndEC Model

The EndEC model is intended to demonstrate how the Repast S platform's support for the Groovy dynamic language can be used to endogenously model the emergence of coordination. The model itself consists of a set of agents each with an individual list of movement capabilities and utterances. Each agent's movement capabilities and utterances are drawn from supersets of capabilities and utterances that are available to the population as a whole. The agents each maintain their own individual association of movements with utterances. As the simulation proceeds, agents make utterances to announce their movements and, over time, adjust their associations to match the observed behavior of their neighbors.

Upon initialization of the EndEC model, each agent is given a random subset of movement capabilities, out of a set of the total capabilities $\{C_1, C_2, \ldots C_{20}\}$ (Fig. 1a). Each agent is also given a random subset of utterances, out of a set of total utterances {'a', 'b', ... 'z'} (Fig. 1b). To each agent utterance is assigned an action, which is randomly composed from the capabilities available to the particular agent (Fig. 1c). Thus, at the start of a simulation, every agent will only know a subset of total utterances and will have their own unique actions associated with each of the utterances.

At each simulation time step, an agent is randomly chosen to be a 'local commander' of the other agents within a certain distance defined to be the agent's 'neighborhood.' The local commander issues a command, as an utterance, to the agents in their neighborhood (Fig. 2a). Some agents will already possess knowledge of the uttered command, others may not (Fig. 2b).

All the agents involved, the commander and those commanded, proceed to execute the uttered command. The agents that already possess knowledge of the command (i.e., those with a movement action associated with the utterance) will execute their associated movement action (Fig. 3a). The agents for which the command is new will compose a new random action from their own capabilities list and associate it with the newly encountered command utterance, and then proceed to execute the new action (Fig. 3b).

After execution (Fig. 4a), each commanded agent compares its own action result with that of the commander (Fig. 4b) and decides whether to and how to modify its

[3]It is perhaps useful to point out that, from a computer science point of view, the ability to endogenously modify code is not a new concept. LISP (one of the first high-level languages) incorporated this ability and was used widely in AI research. Here we present how to apply such dynamic language capabilities to social simulation via agent-based modeling in Repast Simphony.

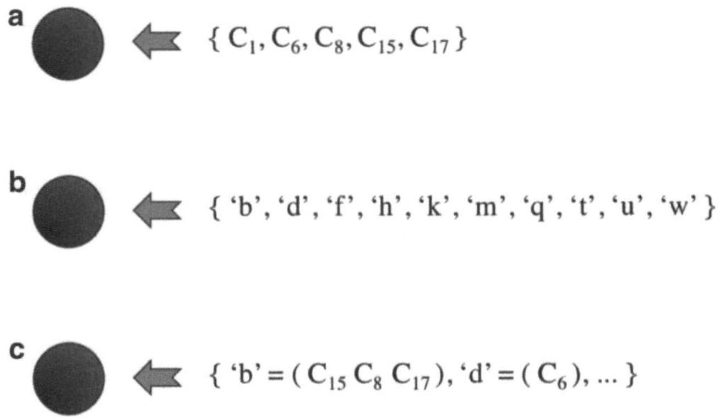

Fig. 1 EndEC model agent initialization. Upon initialization, each agent is (**a**) given a random subset of movement capabilities, out of a set of the total capabilities {C_1, C_2, ... C_{20}} and (**b**) given a random subset of utterances, out of a set of total utterances {‘a’, ‘b’, ... ‘z’}. Then, (**c**) each agent’s utterance is assigned an action, randomly composed from the agent’s available capabilities

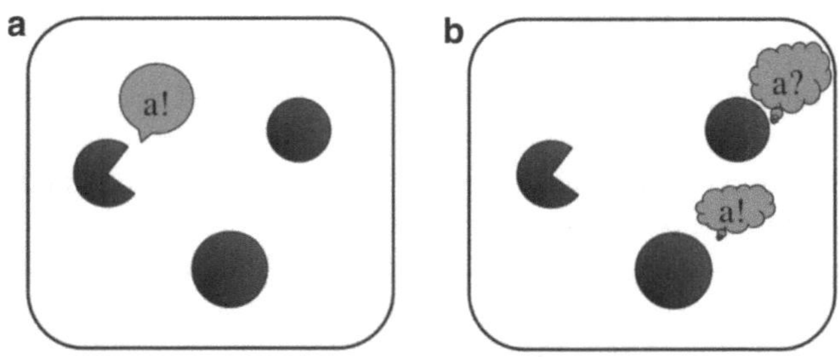

Fig. 2 Commander agent and neighbors. (**a**) A local commander issues a command, as an utterance, to the agents in its neighborhood. (**b**) One agent already possesses knowledge of the uttered command (the agent with the ‘a!’ thought bubble), while the other agent does not

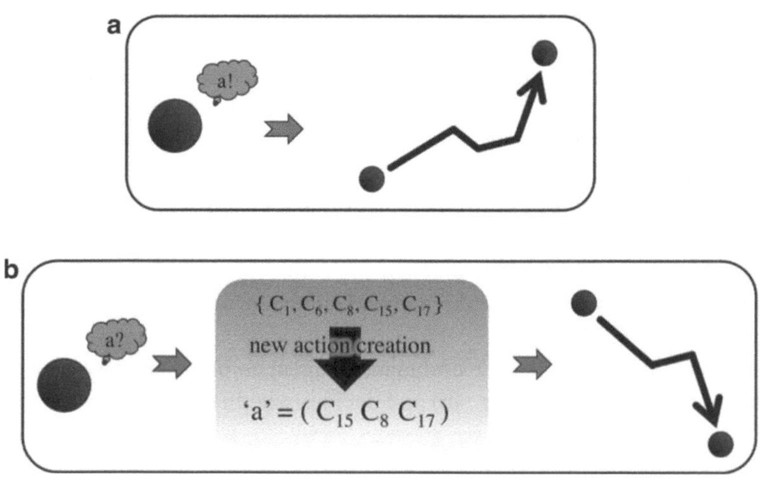

Fig. 3 Agent motion. (**a**) The agents which possess knowledge of the issued command execute their associated movement action. (**b**) The agents for which the command is new compose a new random action from their own capabilities list and associate it with the newly encountered command utterance, and proceed to execute the new action

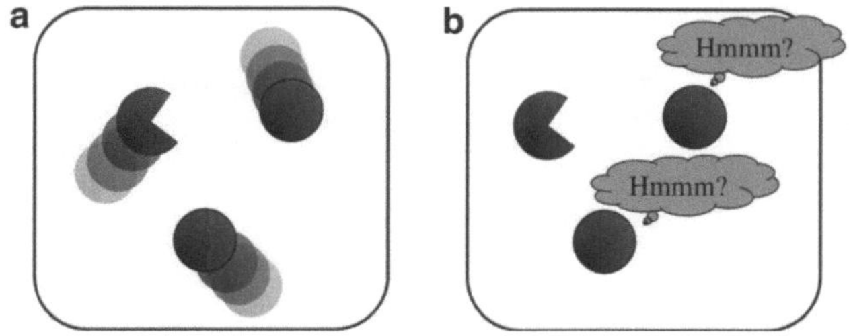

Fig. 4 Movement comparison with commander. (**a**) The agents (commander and commanded) execute their movement actions. (**b**) Each commanded agent compares its own action result with the observed behavior of the commander and determines whether to and how to modify its action such that it corresponds more closely to the observed behavior of the commander

action using the fixed number of capabilities it possesses. It is important to note that the agent's movement capabilities lists are fixed and unique. Therefore, in the general case, each agent must learn to combine different movements to build associations that match their peers. In this way, the agents which start off with different capabilities, vocabularies, and languages, gradually learn each other's utterances and adapt the actions associated with those utterances. It is in this way that the agents are said to endogenously coordinate.

4 Implementation of the EndEC Model

The Repast S implementation of the EndEC model is founded on the use of Groovy closures. According to Steckler and Wand [11] a closure is an independent segment of code that can be bound at will to a set of variables and then executed.

Groovy closures offer a concise and powerful way to encapsulate agent behaviors as first class objects. This enables us to define the total set of possible agent capabilities $\{C_1, C_2, \ldots C_{20}\}$ simply as a list of closures. Figure 5 is a sample code snippet demonstrating this process. We first use the dynamic typing of Groovy to define the empty list **allCapabilitiesClosures**. The **def** keyword is the wildcard type, allowing us to introduce new variables without the need to define a priori the specific type of variable we want. (As shown later, this kind of dynamic typing is optional and we can specify a variable's type whenever we so desire.) We then utilize one of the many convenient Groovy iteration methods **times** to create a number **numOfTotalCapabilities** of **tempClosure** closures. Each **tempClosure** is defined such that, when executed, it calls the **move** method of the agent bound to the **a** parameter.[4] Each **tempClosure** instance, corresponding to one of the

[4]Groovy implements closures using the curly brace syntax, where input parameters are separated from the body of the closure by the '->' character. If input parameters are not specified and the '->' character is omitted a one parameter closure is assumed, where the parameter is accessed via the **it** keyword.

```
def allCapabilityClosures = []
numOfTotalCapabilities.times {
        def coords = getRandomCoordinates()
        def tempClosure = {a ->
             a.move(coords)
        }
        allCapabilityClosures << tempClosure
}
```

Fig. 5 Agent capabilities defined as closures

capabilities in $\{C_1, C_2, \ldots C_{20}\}$, is defined with a different set of random coordinates **coords** being passed to the **move** method. Once a **tempClosure** is defined, it is appended to the **allCapabilityClosures** list.

Each agent is assigned a subset of the closures in **allCapabilityClosures** as its own collection of capabilities. Once such a subset is determined, let's call it **unboundCapabilities**, we 'curry'[5] the closures in this set by binding the first parameter of each closure to an agent reference. That is, the parameter **a** of each capability closure is bound to an agent reference, resulting in a zero parameter closure. In Fig. 6 we demonstrate how we accomplish this, with the implied assumption that the code snippet is part of the agent class and, hence, the **this** keyword refers to a specific agent instance. The collection method **collect** takes a closure as a parameter (parentheses are often optional in Groovy method calls) and this closure operates on each element of the collection **unboundCapabilities**, returning the processed list. Each element of the resulting list **capabilities** is the zero parameter capability closure returned by the currying operation **it.curry(this)**, where **it** refers to an unbound closure from the **unboundCapabilities** list.

As mentioned earlier, at the simulation initialization stage each agent is given a random set of utterances and an action randomly composed from the capabilities available to the particular agent is associated with each utterance. Figure 7 demonstrates the first step in this process, the creation of a list of random length (a single element list to a **maxCapabilitiesComposition** length list) populated by randomly chosen capability closures (chosen from the agent's own capabilities). In Fig. 8 we define a closure which iterates through the list of closures created in Fig. 7 and executes each of them (the **it()** call executes the closure **it**). Finally, in Fig. 9 we add a method (implemented as the closure from Fig. 8) with the same name as the utterance to the agent object. The final step is an important one and warrants further explanation. Since the release of Groovy version 1.5, both methods and fields can be added to classes and instances at runtime via the 'ExpandoMetaClass.' This powerful meta-programming ability for a model to be implemented using a fully object-oriented language while still allowing the agents to fundamentally change their structures (and model objects in general) as the simulation executes brings a

[5]'Currying' is the process of assigning values to some or all of the free parameters of a closure.

```
capabilities = unboundCapabilities.collect {
    it.curry(this)
}
```

Fig. 6 Currying closures

```
def randomClosureList = []
RandomHelper.nextIntFromTo(1,maxCapabilitiesComposition)
    .times{
        randomClosureList << capabilities[
        RandomHelper.nextIntFromTo(0,maxCapabilities-1)]
}
```

Fig. 7 Creating a random length list of closures from a random selection of agent capabilities

```
def closure = {
        randomClosureList.each{it()}
}
```

Fig. 8 Creating a closure from a list of closures

```
void addMethod(String methodName, methodClosure){
    def emc
    if (this.metaClass instanceof ExpandoMetaClass){
        emc = this.metaClass
    }
    else{
        emc = new ExpandoMetaClass(this.class, false, true)
    }
    emc."${methodName}" = methodClosure
    emc.initialize()
    this.metaClass = emc
}
```

Fig. 9 Adding a method to a specific agent via the ExpandoMetaClass in Groovy 1.5

whole new open-ended level of flexibility to agent based modeling! [6] To review the progress so far, Figs. 5–9 described how we implemented the EndEC agent initialization (described in Fig. 1) using Groovy techniques.

Once the simulation is started, at each time step, a randomly chosen agent (the commander) issues a command to agents in its neighborhood (see Fig. 2). The implementation of this process is demonstrated in Fig. 10. The **ContinuousWithin** query object we create is a Repast Simphony class and its **query** method returns an iterator over the neighbors within the **neighborhoodDistance**. We use a Groovy **for**-loop to scan across the neighbors and we issue commands by invoking the

[6] Groovy 1.6 introduced additional meta-programming capabilities, including the ExpandoMetaClass Domain Specific Language and per instance meta-classes for Java objects, making meta-programming even more powerful and convenient.

```
def query = new ContinuousWithin(space, this,
      neighborhoodDistance)
def nearbyAgents = query.query()
for (neighbor in nearbyAgents){
      neighbor."do${command}"()
}
```

Fig. 10 Finding nearby agents and issuing a command

```
def methodMissing(String name, args) {
    String methodName = name - "do"
    if (this.metaClass.respondsTo(this, methodName)){
    }
    else {
          utterancesMap."${methodName}" = createRandomClosureList()
          addMethod(methodName,
                createActionClosureFromClosureList(
                      utterancesMap."${methodName}"))
    }
    this."${methodName}"()
}
```

Fig. 11 The methodMissing agent method

neighbors' methods. Each agent method call is constructed via a Groovy 'GString.' GStrings allow one to embed code within a string, where double quotes are used to differentiate between GString objects and regular String objects (single quotes). For clarification, if, for example, the **command** utterance issued by the commander is '**a**', the resulting agent method call will be **neighbor.doa()**.

However, if we refer back to Fig. 9, where the agent utterance methods were defined, we notice that the agent method names are simply the utterance names. Thus, by calling **neighbor.doa()** (instead of **neighbor.a()**) we are referencing a nonexistent agent method. In Java this would result in a compiler error but in Groovy, we can successfully make calls to nonexistent methods as long as we define a **methodMissing** method in the called object's class. The **methodMissing** method intercepts any calls to nonexistent class methods. Our implementation of **methodMissing** is in Fig. 11. We first strip the "**do**" string from the method call, leaving only the utterance part of the command. Then, if the object called responds to the resulting method call (i.e., if it has the method associated with the utterance), the method is called (Fig. 5a). On the other hand, if after the "**do**" is stripped from the method call the agent is found not to have a method associated with the utterance, a new random method with the name of the utterance is created with a call to **addMethod** (see Fig. 3b). After the new method is created, it is called.

One of the most important features of Groovy is its ability to smoothly integrate with Java, allowing one to leverage any existing Java libraries without any rewriting of code. This is what makes Groovy integration with Repast S possible. Issues can potentially arise, however, when one has to make calls to Java methods (which expect typed parameters) using dynamically typed variables.

```
public void move(displacement){
  Context context = ContextUtils.getContext(this)
  ContinuousSpace space = (ContinuousSpace) context
        .getProjection("Space")
  space.moveByDisplacement(this, displacement as double[])
}
```

Fig. 12 An example of dynamic typing and coercion

```
switch(candidateMapEntry.key){
case 'keep':
    // 'keep' code
    break
case 'pop':
    // 'pop' code
    break
case ~/add.*/:
    // 'add' code
    break
default:
    // 'default' code
    break
}
```

Fig. 13 Regular expressions in Groovy switch statements

Fortunately, Groovy's automatic type coercion makes this process painless. In Fig. 12 (corresponding to Fig. 4a in the EndEC model description section) we have defined a method in Groovy which takes a dynamically typed parameter **displacement**. When we call the Java method **space.moveByDisplacement** which requires a type **double[]** second parameter, we simply use the **as** keyword to coerce the displacement variable to the **double[]** type. Of course, such type coercion can only succeed when conversion from one type to the other is well-defined.

Finally, an additional Groovy feature that greatly enhances the ability to write concise yet expressive code is the **switch** statement. One of the reasons why the Groovy **switch** statement is more flexible than its Java counterpart is its ability to optionally use regular expressions. Figure 13 demonstrates how we utilize the **switch** statement in the EndEC model (this corresponds to Fig. 4b, the EndEC agent learning process). The third **case** is satisfied by any string starting with "**add**". So, instead of writing out all the possible exact matches, we concisely specify the **case** as a regular expression pattern.

Figure 14 is a visualization of the EndEC model within Repast S. The process of utterance learning by EndEC agents is depicted in Fig. 15a, where, through the course of a typical simulation run, a rise in mean recognized utterances per agent and a gradual decline in the standard deviation of utterances per agent is observed. In Fig. 15b, we present evidence of the process of agent action adaptation, via the

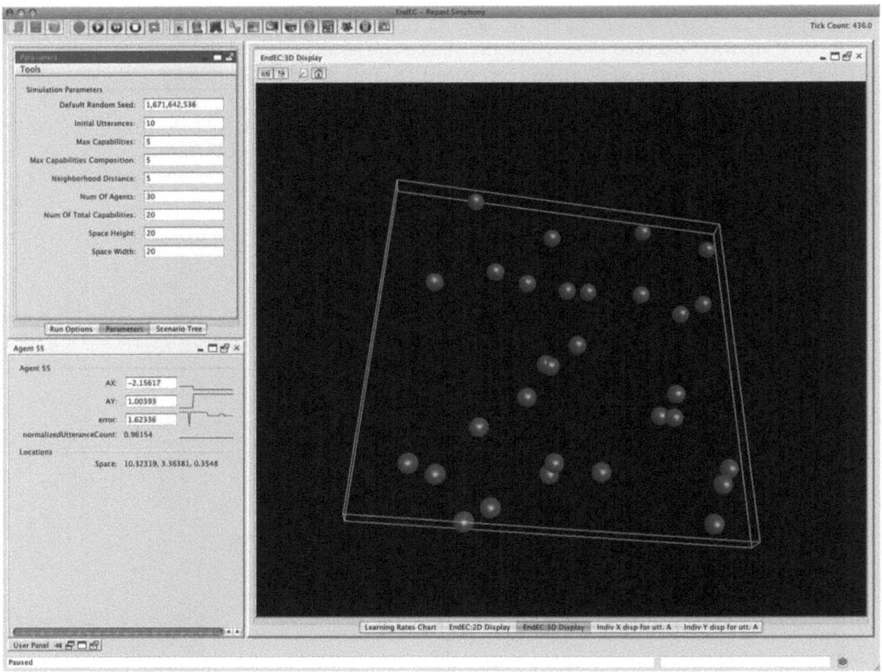

Fig. 14 The EndEC model as implemented in Repast S

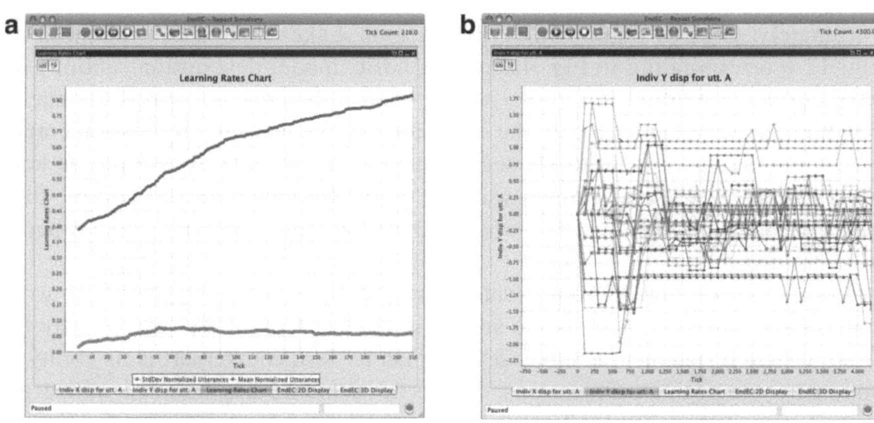

Fig. 15 Illustrative EndEC model outputs. (**a**) A rise in the mean recognized utterances per agent and a gradual decline in the standard deviation of utterances per agent. (**b**) The evolution of the y-components of the action associated with utterance ‘a’ for all agents

evolution of all the y-components of the actions associated with utterance ‘a’ for all individual agents. The values are seen to only weakly converge, reflecting the fact that, since each agent’s movement capabilities list is fixed and unique, it can do only so much to mimic the movement of its peers. As an additional example,

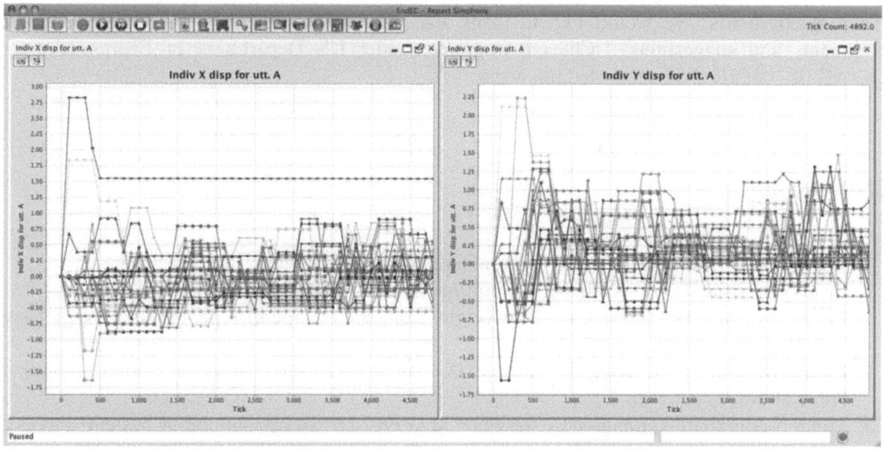

Fig. 16 Additional illustrative EndEC model outputs. A comparison of the x (*left*) and y (*right*) components of the action associated with utterance 'a' for all agents

Fig. 16 shows a cross comparison of the x and y components of the 'a' utterance action for all agents.

5 Conclusions

This paper presented a model of the endogenous emergence of coordination and highlighted the ways in which a dynamic language, in this case Groovy, was found to be particularly helpful during model implementation. The EndEC model agents were shown to acquire new utterances and modify their understanding of their utterances as the simulation progressed. While it would have been possible to implement the EndEC model with a static language (e.g., Java), we would have lost a lot in terms of the clarity of the data structures used and the speed with which we developed the model. The EndEC model is a relatively simple model mainly developed to showcase the promise of dynamic language capabilities in social simulation, but it is quite possible that it will see further development in the future. As such, the flexibility and intuitiveness of the dynamic aspects of the model would be crucial for scaling up or adding further complexity. Ultimately what is particularly exciting is the potential for creating truly dynamic and evolving open-ended simulations, where the simulation fundamentally changes as it executes. The phenomenon of second-order emergence, where structures not present at the start of a simulation not only emerge but become 'real' in the minds and decision making of agents (e.g., the emergence of groups), is an interesting and important research area where the dynamic evolution of simulation structures afforded by languages like Groovy could prove very helpful.

Acknowledgments We would like to thank the anonymous reviewers who offered helpful comments and suggestions. UChicago Argonne, under US Department of Energy contract DE-AC-02- 06CH11357 supported this work.

References

1. Apple (2007) The objective-C programming language
2. Groovy Website (2009) Available at: http://groovy.codehaus.org/
3. Ke J, Holland JH (2006) Language origin from an emergentist perspective. Appl Ling 27:691–716
4. König D, Glover A, King P, Laforge G, Skeet J (2007) Groovy in action. Manning Publications, Greenwich
5. Minar N, Burkhart R, Langton C, Askenazi M (1996) The swarm simulation system: a toolkit for building multi-agent simulations. In: Santa Fe Institute Working Paper 96-06-042
6. North MJ, Howe TR, Collier NT, Vos JR (2005) The repast simphony development environment. In: Macal CM, North MJ, Sallach DL (eds) Proceedings of the agent 2005 conference on generative social processes, models, and mechanisms, Chicago, IL, USA, 13–15 Oct 2005
7. North MJ, Howe TR, Collier NT, Vos JR (2005) The repast simphony runtime system. In: Macal CM, North MJ, Sallach DL (eds) Proceedings of the agent 2005 conference on generative social processes, models, and mechanisms, Chicago, IL, USA, 13–15 Oct 2005
8. Ozik J, North MJ (2008) Agent-based modeling with a dynamic language: platform support for modeling endogenous coordination. In: Proceedings of the second world congress on social simulation, George Mason University, Fairfax, VA USA
9. Ozik J, North MJ, Sallach DL, Panici JW (2007) ROAD map: transforming and extending repast with groovy. In: Macal CM, North MJ, Sallach DL (eds) Proceedings of the agent 2007 conference on complex interaction and social emergence, Evanston, IL, USA, 15–17 Nov 2007
10. ROAD, Repast, Repast Organization for Architecture and Design Home Page, Chicago, IL USA (2008). Available at: http://repast.sourceforge.net
11. Steckler PA, Wand M (1997) Lightweight closure conversion. ACM Trans Program Lang Syst 19(1):48–86
12. Swarm Development Group (2008) SDG home page. Available at: http://www.swarm.org/
13. Tao G, Minett JW, Jinyun K, Holland JH, Wang WSY (2005) Coevolution of lexicon and syntax from a simulation perspective: research articles. Complex 10:50–62

Author Index

Keyword Index

MIX
Papier aus verantwortungsvollen Quellen
Paper from responsible sources
FSC® C105338

If you have any concerns about our products,
you can contact us on
ProductSafety@springernature.com

In case Publisher is established outside the EU,
the EU authorized representative is:
Springer Nature Customer Service Center GmbH
Europaplatz 3, 69115 Heidelberg, Germany

Printed by Libri Plureos GmbH
in Hamburg, Germany